AF597771

Springer Theses

Recognizing Outstanding Ph.D. Research

For further volumes:
http://www.springer.com/series/8790

Aims and Scope

The series "Springer Theses" brings together a selection of the very best Ph.D. theses from around the world and across the physical sciences. Nominated and endorsed by two recognized specialists, each published volume has been selected for its scientific excellence and the high impact of its contents for the pertinent field of research. For greater accessibility to non-specialists, the published versions include an extended introduction, as well as a foreword by the student's supervisor explaining the special relevance of the work for the field. As a whole, the series will provide a valuable resource both for newcomers to the research fields described, and for other scientists seeking detailed background information on special questions. Finally, it provides an accredited documentation of the valuable contributions made by today's younger generation of scientists.

Theses are accepted into the series by invited nomination only and must fulfill all of the following criteria

- They must be written in good English.
- The topic should fall within the confines of Chemistry, Physics, Earth Sciences, Engineering and related interdisciplinary fields such as Materials, Nanoscience, Chemical Engineering, Complex Systems and Biophysics.
- The work reported in the thesis must represent a significant scientific advance.
- If the thesis includes previously published material, permission to reproduce this must be gained from the respective copyright holder.
- They must have been examined and passed during the 12 months prior to nomination.
- Each thesis should include a foreword by the supervisor outlining the significance of its content.
- The theses should have a clearly defined structure including an introduction accessible to scientists not expert in that particular field.

Xu Hou

Bio-inspired Asymmetric Design and Building of Biomimetic Smart Single Nanochannels

Doctoral Thesis accepted by
Chinese Academy of Sciences, China

Springer

Author
Dr. Xu Hou
National Center for Nanoscience and Technology
Beijing
People's Republic of China

Present Address
School of Engineering and Applied Sciences
Wyss Institute for Biologically Inspired Engineering at Harvard University
Cambridge, MA
USA

Supervisor
Prof. Lei Jiang
Institute of Chemistry
Chinese Academy of Sciences
Beijing
People's Republic of China

ISSN 2190-5053 ISSN 2190-5061 (electronic)
ISBN 978-3-662-51288-3 ISBN 978-3-642-38050-1 (eBook)
DOI 10.1007/978-3-642-38050-1
Springer Heidelberg New York Dordrecht London

© Springer-Verlag Berlin Heidelberg 2013
Softcover reprint of the hardcover 1st edition 2013
This work is subject to copyright. All rights are reserved by the Publisher, whether the whole or part of the material is concerned, specifically the rights of translation, reprinting, reuse of illustrations, recitation, broadcasting, reproduction on microfilms or in any other physical way, and transmission or information storage and retrieval, electronic adaptation, computer software, or by similar or dissimilar methodology now known or hereafter developed. Exempted from this legal reservation are brief excerpts in connection with reviews or scholarly analysis or material supplied specifically for the purpose of being entered and executed on a computer system, for exclusive use by the purchaser of the work. Duplication of this publication or parts thereof is permitted only under the provisions of the Copyright Law of the Publisher's location, in its current version, and permission for use must always be obtained from Springer. Permissions for use may be obtained through RightsLink at the Copyright Clearance Center. Violations are liable to prosecution under the respective Copyright Law.
The use of general descriptive names, registered names, trademarks, service marks, etc. in this publication does not imply, even in the absence of a specific statement, that such names are exempt from the relevant protective laws and regulations and therefore free for general use.
While the advice and information in this book are believed to be true and accurate at the date of publication, neither the authors nor the editors nor the publisher can accept any legal responsibility for any errors or omissions that may be made. The publisher makes no warranty, express or implied, with respect to the material contained herein.

Printed on acid-free paper

Springer is part of Springer Science+Business Media (www.springer.com)

Parts of this thesis have been published In the following journal articles:

Hou X, Zhang HC, Jiang L (2012) Building Bio-inspired Artificial Functional Nanochannels: From Symmetric to Asymmetric Modification. Angewandte Chemie International Edition 51 (22):5296–5307. doi: 10.1002/anie.201104904

Hou X, Guo W, Jiang L (2011) Biomimetic Smart Nanopores and Nanochannels. Chemical Society Reviews 40 (5):2385–2401. doi: 10.1039/C0cs00053a

Hou X, Yang F, Li L, Song YL, Jiang L, Zhu DB (2010) A Biomimetic Asymmetric Responsive Single Nanochannel. Journal of the American Chemical Society 132 (33):11736–11742. doi: 10.1021/Ja1045082

Hou X, Liu YJ, Dong H, Yang F, Li L, Jiang L (2010) A pH-Gating Ionic Transport Nanodevice: Asymmetric Chemical Modification of Single Nanochannels. Advanced Materials 22 (22):2440–2443. doi: 10.1002/adma.200904268

Hou X, Dong H, Zhu DB, Jiang L (2010) Fabrication of Stable Single Nanochannels with Controllable Ionic Rectification. Small 6 (3):361–365. doi: 10.1002/smll.200901701

Hou X, Jiang L (2009) Learning from Nature: Building Bio-inspired Smart Nanochannels. Acs Nano 3 (11):3339–3342. doi: 10.1021/Nn901402b

Hou X, Guo W, Xia F, Nie FQ, Dong H, Tian Y, Wen LP, Wang L, Cao LX, Yang Y, Xue JM, Song YL, Wang YG, Liu DS, Jiang L (2009) A Biomimetic Potassium Responsive Nanochannel: G-Quadruplex DNA Conformational Switching in a Synthetic Nanopore. Journal of the American Chemical Society 131 (22):7800–7805. doi: 10.1021/Ja901574c

Zhang MH, Hou X, Wang JT, Tian Y, Fan X, Zhai J, Jiang L (2012) Light and pH Cooperative Nanofluidic Diode Using a Spiropyran-Functionalized Single Nanochannel. Advanced Materials 24 (18):2424–2428. doi: 10.1002/adma.201104536

Zhang QQ, Liu ZY, Hou X, Fan X, Zhai J, Jiang L (2012) Light-Regulated Ion Transport Through Artificial Ion Channels Based on TiO_2 Nanotubular Arrays. Chemical Communications 48 (47):5901–5903. doi: 10.1039/C2cc32451b

Tian Y, Wen LP, Hou X, Hou GL, Jiang L (2012) Bio-inspired Ion-Transport Properties of Solid-State Single Nanochannels and Their Applications in Sensing. ChemPhysChem 13 (10):2455–2470. doi: 10.1002/cphc.201200057

Han CP, Hou X, Zhang HC, Guo W, Li HB, Jiang L (2011) Enantioselective Recognition in Biomimetic Single Artificial Nanochannels. Journal of the American Chemical Society 133 (20):7644–7647. doi: 10.1021/Ja2004939

Tian Y, Hou X, Jiang L (2011) Biomimetic Ionic Rectifier Systems: Asymmetric Modification of Single Nanochannels by Ion Sputtering Technology. Journal of Electroanalytical Chemistry 656 (1-2):231–236. doi: 10.1016/j.jelechem.2010.11.005

Dong H, Nie RX, Hou X, Wang PR, Yue JC, Jiang L (2011) Assembly of F0F1-ATPase into Solid State Nanoporous Membrane. Chemical Communications 47 (11):3102–3104. doi: 10.1039/C0cc05107a

Wen LP, Hou X, Tian Y, Nie FQ, Song YL, Zhai J, Jiang L (2010) Bio-inspired Smart Gating of Nanochannels Toward Photoelectric-Conversion Systems. Advanced Materials 22 (9):1021–1024. doi: 10.1002/adma.200903161

Tian Y, Hou X, Wen LP, Guo W, Song YL, Sun HZ, Wang YG, Jiang L, Zhu DB (2010) A Biomimetic Zinc Activated Ion Channel. Chemical Communications 46 (10):1682–1684. doi: 10.1039/B918006k

Wen LP, Hou X, Tian Y, Zhai J, Jiang L (2010) Bio-inspired Photoelectric Conversion Based on Smart-Gating Nanochannels. Advanced Functional Materials 20 (16):2636–2642. doi: 10.1002/adfm.201000239

Guo W, Xia HW, Xia F, Hou X, Cao LX, Wang L, Xue JM, Zhang GZ, Song YL, Zhu DB, Wang YG, Jiang L (2010) Current Rectification in Temperature-Responsive Single Nanopores. ChemPhysChem 11 (4):859–864. doi: 10.1002/cphc.200900989

Xia F, Guo W, Mao YD, Hou X, Xue JM, Xia HW, Wang L, Song YL, Ji H, Qi OY, Wang YG, Jiang L (2008) Gating of Single Synthetic Nanopores by Proton-Driven DNA Molecular Motors. Journal of the American Chemical Society 130 (26):8345–8350. doi: 10.1021/Ja800266p

Supervisor's Foreword

Bio-inspired smart materials should be a "live" material with various functions like living organism in Nature. Ion channels that exist in living organisms play important roles in maintaining normal physiological conditions and serve as "smart" gates to ensure selective ion transport. Normal body function depends strongly on regulation of ion transport inside these nanochannels. Thus, designing a system that simulates these complex processes in living systems is a challenging task for nanoscience. Fabrication and application of artificial nanochannels are becoming the focus of attention because, compared with their biological counterparts, they offer greater flexibility in terms of shape and size, superior robustness, and surface properties that can be tuned depending on the desired function. Chemical modification of the interior surface of the nanochannels with functional molecules that closely mimic the gating mechanisms of biological channels may provide a highly efficient means to control ionic or molecular transport through nanometer-scale openings in response to ambient stimuli, such as temperature, light, pH, and specific ions. In this thesis, a comprehensive and systematic strategy for the design and preparation of artificial responsive symmetric/asymmetric nanochannel systems under various symmetric/asymmetric stimuli is presented for the first time. It is intended to utilize ion-track-etching polymer nanochannels with different shapes as examples to demonstrate the feasibility of the design strategy of building novel artificial functional nanochannels by various symmetric/asymmetric physicochemical modifications.

Xu had been working on biomimetic intelligent nanomaterials research for about four years. His scientific interests are focused on the design and fabrication of biomimetic smart single nanochannel materials. I am truly convinced that he is an excellent example of a highly skilled researcher, working alongside chemists, physicists, engineers, and biologists, he does innovative and collaborative research in the emerging field of interdisciplinary nano/bio science. His research carries the hallmarks of brilliance and creativity of a quality and in a quantity that is rare to behold at the doctoral level. Xu is a diligent man, ambitious and eager to face the challenges in his research fields. It is truly amazing to examine the amount and quality of research work he has completed in such a brief period. Many research groups all over the world have cited his papers as the first reference in their research papers, such as Max-Planck-Institute, Germany; Pennsylvania State University, USA; University of Southampton, U.K.; and Catalan Institute of Nanotechnology, Spain. In my contact with him in my lab, his creativity, hard work, and attitude to research work gave me a deep impression. He always proposed new ideas on his research and could resolve problems by himself, and I was very satisfied with him. He was also very friendly and cooperative in my group. I believe that he will achieve greater academic success in his future work, and this thesis will be of immediate interest to a broad readership.

Beijing, February 2013 Lei Jiang

Preface

I believe a number of exciting developments in the near future may be anticipated from the design, preparation process, and experimental results reported in this thesis. The increase in relevant publications in this field clearly indicates that the design and synthesis of artificial responsive nanochannel materials offer a flexible venue to create various biomimetic intelligent apparatus. It is an emerging field that in many respects is still in its infancy. With the inspiration of examples from nature, our scientific community started to build up various biomimetic smart nanochannels. The ability to tune the shape and the surface chemical properties of the nanochannel materials affords a flexible venue to address a host of questions and problems at the challenging forefront of nanotechnology and materials science. In order to pursue 'smart,' further ambient stimuli responsive materials need to take into account properties such as magnetic response, acoustic response, and integrated various stimuli response. In addition, making smart nanochannels stable, reversible, and durable, it will remain essential for the successful implementation of the expected practical applications for real-world applications. For future progress, it will be important to further improve the fabrication for various shapes of the nanochannels and, more importantly, to maximize efforts to create more ambitious smart functional molecules to be immobilized onto the inner nanochannels.

Here, I would like to express my sincere appreciation to my advisor Prof. Lei Jiang for his encouragement and powerful support. I also thank Prof. Daoben Zhu, Prof. Yanlin Song, Prof. Wenping Hu, Prof. Lin Li, Prof. Dongsheng Liu, Prof. Shu Wang, Prof. Shutao Wang, Prof. Yong Zhao, Prof. Yugang Wang, Prof. Zhongfan Liu, Prof. Dong Han, Prof. Zhiyong Tang, Prof. Yuliang Zhao, Prof. Xi Zhang, Prof. Jin Zhai et al., and Dr. Fan Xia, Dr. Wei Guo, Dr. Fu Yang, Dr. Hua Dong, Dr. Huacheng Zhang, Dr. Ye Tian, Dr. Cuiping Han, Ms. Zeng Lu, Ms. Yujie Liu, Mr. Zhe Yang et al. Thank you very much. The research work in this thesis was supported by the National Research Fund for Fundamental Key Projects (2011CB935703, 2010CB934700, 2009CB930404, 2007CB936403), and

National Natural Science Foundation (20974113, 20920102036). The Chinese Academy of Sciences is gratefully acknowledged. Lastly, my thanks go to my beloved family for their loving consideration and great confidence in me all through these years.

Cambridge, MA, March 2013 Xu Hou

Contents

Chapter 1
Introduction

Over millions of years, complex processes of intelligent control have been evolved in nature. Structures from nature have remarkable properties, many of which have inspired laboratory research. Bio-inspired materials and devices are attracting increasing interest because of their unique properties, which have paved the way to many significant applications [1–3]. For example, ion channels play a very important role in basic biochemical processes and maintaining normal physiological conditions in cells. In this chapter, it is intended to utilize a specific responsive behavior as an example to demonstrate the feasibility of various design strategies for building bio-inspired artificial functional nanochannels. This specific responsive behavior is to regulate ionic transport properties inside the single nanopore or nanochannel. It is also intended to provide an overview of this fascinating research filed.

1.1 Ion Channels

Nature provides a huge range of biological materials, with various smart functions over millions of years of evolution, just as ion channels, which serve as a large source of bio-inspiration for biomimetic materials. The transport of ions across cell membranes is a prerequisite for many of life's processes. Life depends on the continued flow of ions into and out of the cells.

Cells use biological nanochannels including ion pumps and ion channels, embedded within cell membranes, to communicate chemically and electrically with the extracellular world. Ion channels that open and close in response to ambient stimuli for regulating ion permeation through cell membranes are significant for the implementation of various important physiological functions in life processes (Fig. 1.1).

Ion channels are membrane protein complexes. The complex proteins with various responsive conformational changes regulate ionic transport properties of ion channels. Regulation of ionic transport is mainly composed of three

X. Hou, *Bio-inspired Asymmetric Design and Building of Biomimetic Smart Single Nanochannels*, Springer Theses,
DOI: 10.1007/978-3-642-38050-1_1, © Springer-Verlag Berlin Heidelberg 2013

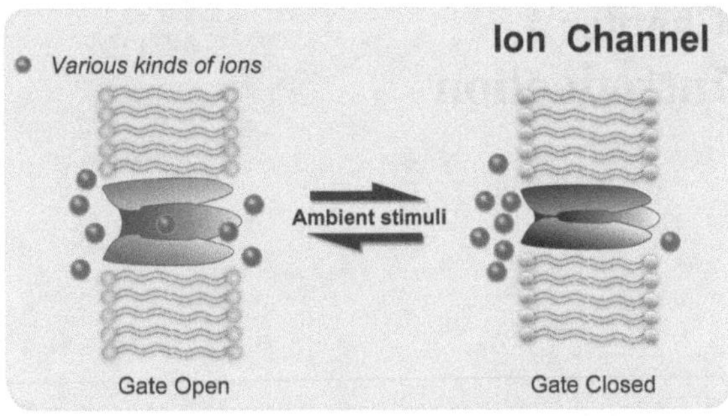

Fig. 1.1 Schematic representation of a biological ion channel open and close in response to ambient stimuli

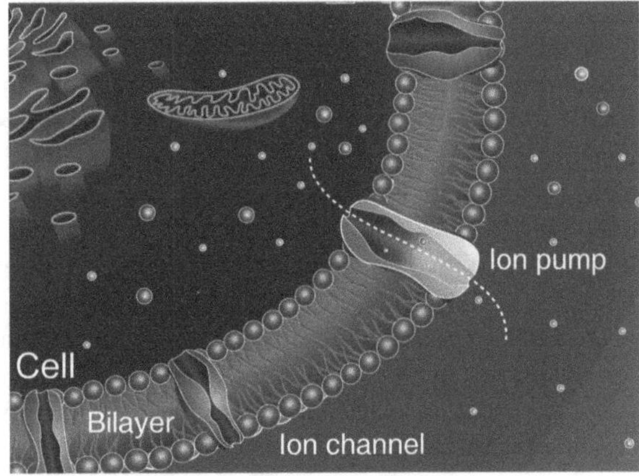

Fig. 1.2 Different ion channels and ion pumps in cell

characteristic features including ionic selectivity, ionic rectification and ionic gating. Various components of ion channels are not uniform in distribution, and their structures are also asymmetry (Fig. 1.2) [4]. The bio-inspired study of design and development of new artificial biomimetic channels has been receiving a great deal of attention [5–11]. Artificial functional nanochannels have emerged as possible candidates for mimicking physiological processes in biological nanochannels. Moreover, it boosts the development of bio-inspired intelligent nanomachines for real-world applications, such as biosensors, nanofluidic devices and molecular filtration [12–16].

1.2 Design Philosophy of Biomimetic Smart Nanopores/Nanochannels

1.2.1 Definition of Nanopore/Nanochannel

Both "nanopore" and "nanochannel" are in common mutual use [17–25]. From a shape conceptual point-of-view, "nanopore" is defined simply as pore having diameter of 1 to 100 nm, with the pore diameter larger than its depth. If the pore depth is much larger than the diameter, the structure is generally referred to as "nanochannel". For the study on a single nanopore/nanochannel and multiple nanopores/nanochannels, a single one presents an optimal system for studying transport properties of different ions or molecules because one can observe directly the behavior of a single pore or channel without having to average the effects of multiple pores or channels. Nonetheless, multiple nanopores/nanochannels are also very important for the fundamental research. With the improvement of fabrication technology in uniform shape and the homogeneous chemical composition of the nanopores and nanochannels, multiple nanopores/nanochannels will certainly have advantages for achieving large-scale real-world applications.

1.2.2 Classification of Materials for Preparing Nanopores/Nanochannels

According to the specific application requirements, various materials can be selected to prepare artificial nanopores and nanochannels, such as biological (Fig. 1.3a), inorganic (Fig. 1.3b–e), organic (Fig. 1.3g), and composite materials (Fig. 1.3h, i).

Generally, biological materials, due to the stability of the lipid membrane, are subject to certain restrictions [26–28]. Inorganic materials, such as silicon nitride materials (Fig. 1.3d) for Nanopore-DNA sequencing technology have attracted widespread interest [29]. Organic materials, such as polymer films Polyethylene terephthalate (PET) and Polyimide (PI), etc. can be used for preparing various shapes of nanochannels by using the track-etching technique (Fig. 1.3g) [30]. It is worth mentioning that composite materials have the possibility to combine advantages of various materials. For example, metal-polymer asymmetric nanochannels (Fig. 1.3h) show stable and chemically/structurally asymmetric properties for further asymmetric chemical modification with complicated functional molecules for the exploitation of complex smart nanochannel systems [31]. And hybrid biological/solid-state nanopores (Fig. 1.3i) display various potential applications for biotechnology [32, 33].

1.2.3 Design of Biomimetic Smart Nanopores/Nanochannels

As shown in Fig. 1.4, I suggest two routes for the design and preparation of biomimetic smart nanopore/nanochannel materials, which produce biomimetic smart materials for various potential applications. After material selection, different shapes and structures of nanopores or nanochannels were obtained by using various fabrication technologies (Fig. 1.3), such as biological molecules

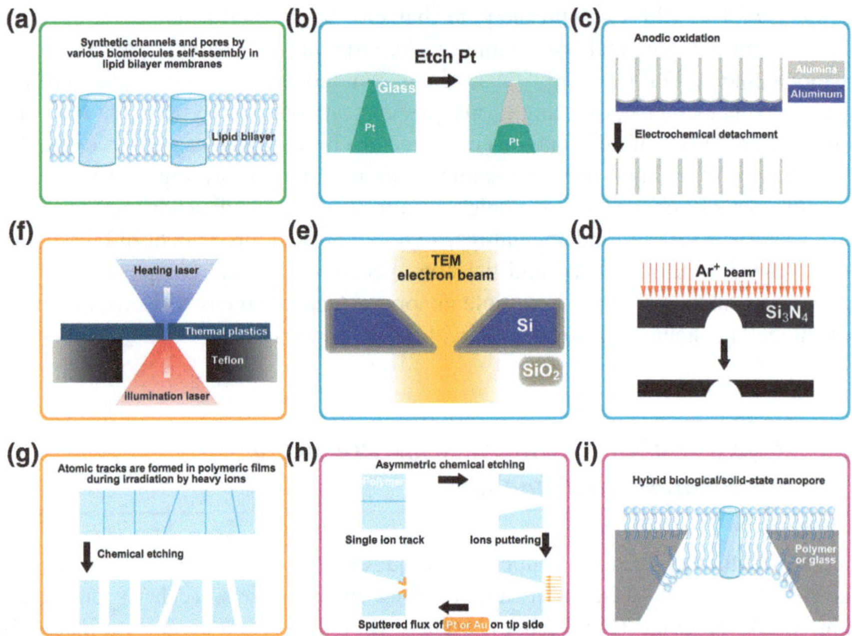

Fig. 1.3 Selection and synthesis of various shapes and structures of nanopore/nanochannel materials; biological materials: **a** artificial ion channels and pores in a lipid bilayer; inorganic materials: **b** asymmetric nanopores prepared by electrochemical etching; **c** multiple nanopores/nanochannels in anodic alumina membranes; **d** nanopores fabricated by ion beam sculpting of Si_3N_4 membranes; **e** nanopores prepared by using electron beam (TEM) technology and anisotropic etching; organic materials: **f** nanopores fabricated by heat induced shrinking of micron scale pores in thermoplastic materials; **g** single or multiple nanochannels prepared by chemical etching of single or multi-track polymer membranes; composite materials: **h** metal-polymer asymmetric nanochannels fabricated by ion-track-etching technique and ion sputtering technology; **i** hybrid biological/solid-state nanopores combined with polymer nanochannels or glass nanopores and biological pores self-assembly in lipid bilayers. Reproduced from Ref. [9] by permission of The Royal Society of Chemistry

self-assembly [26–28], electrochemical etching [34], anodic oxidation method [35], electron beam technology [5, 36], laser technology [37] and ion-track-etching technology [30, 38]. These fabrication approaches have been described in detail, e.g., by Gyurcsanyi [39], Matile et al. [27, 40], Dekker [6], and Martin et al. [41].

Compared with the second route, the first route contains one more step: chemical modification. If the selected material itself has ambient stimuli responsive properties, the nanopore/nanochannel prepared by it will have the same characteristics. For instance, tunable elastomeric nanochannels are compressed to change their cross-sectional size [42]. Therefore, it is an easy way to achieve a simple responsive property by directly using responsive materials for building the nanopores/nanochannels. However, for more complex responsive properties, this route is limited by the availability of suitable materials. It is obvious that chemical

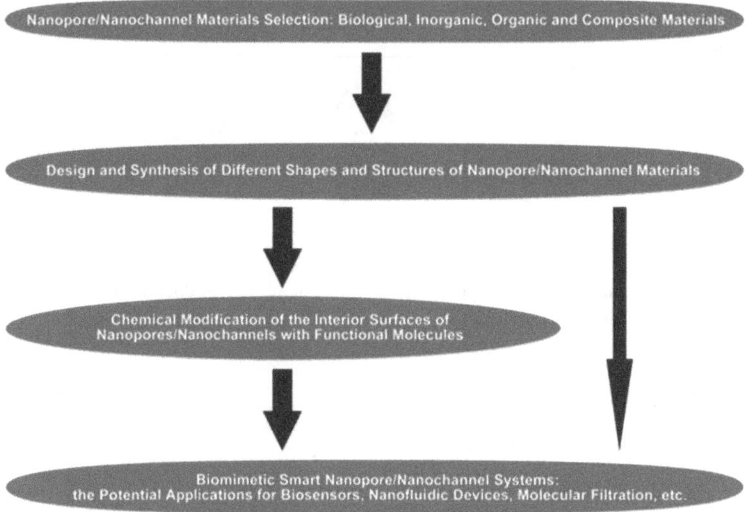

Fig. 1.4 The design and preparation process of biomimetic smart nanopore/nanochannel materials. Reproduced from Ref. [9] by permission of The Royal Society of Chemistry

modification of the interior surfaces of nanopores or nanochannels, which confers much flexibility in changing their physicochemical properties, constitutes a primary approach to make nanopores/nanochannels "smart".

1.2.3.1 Shape Design of Nanopores/Nanochannels

Various shapes of artificial nanopores/nanochannels have been well developed by using specific fabrication techniques, such as ion-track-etching technology, laser technology and electron beam technology. However, these preparation techniques have their limitations which are often associated with the properties of materials and the requirement of size ranges. Therefore, the usual approach is to select materials and preparation methods based on the required shape of nanopores/nanochannels, and then these nanopores/nanochannels can obtain various functional properties through inner surface modification. To note that in this chapter, the shape design is primarily referred to the three-dimensional shape design of nanochannels. While the shape design of nanopores is normally focused on the two-dimensional shape of the pores, because its aspect ratio is small, thus the shape design is similar to the pore-shape design.

Here the examples of ion-track-etched polymer nanochannels with different shapes (Fig. 1.5) would be used to demonstrate the feasibility of the shape design for building functional nanochannels by various symmetric and asymmetric modifications, which may also be extended to other materials.

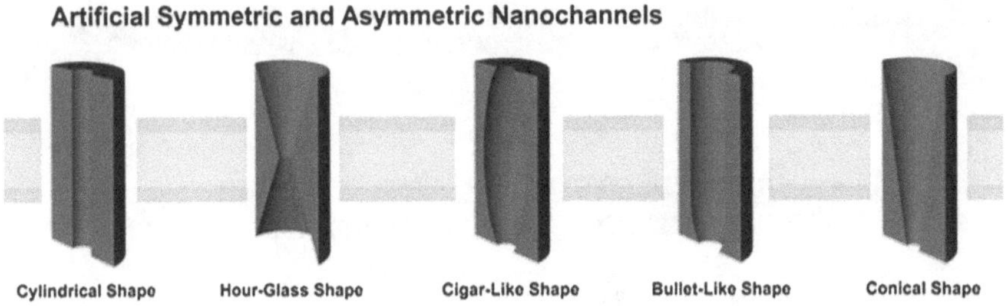

Artificial Symmetric and Asymmetric Nanochannels

Cylindrical Shape Hour-Glass Shape Cigar-Like Shape Bullet-Like Shape Conical Shape

Fig. 1.5 Different shapes of the polymer nanochannels. Reproduced from Ref. [47] by permission of John Wiley & Sons Ltd

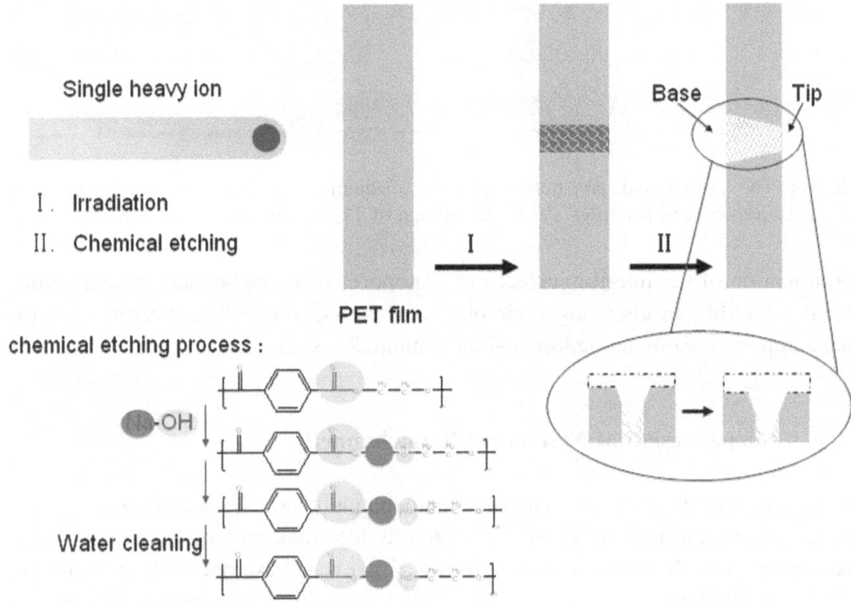

Fig. 1.6 Principle of ion-track-etching technique and the etching process of PET in aqueous NaOH solution. Reprinted with the permission from Ref. [25]. Copyright 2009 American Chemical Society

Ion-track-etching technology, which has become a well-established technology to create very uniform channel in insulator, is based on the following process (Fig. 1.6). When a swift heavy ion passes through the film, it deposits its energy along its trajectory, thus creating a cylindrical damage zone, i.e., the latent track. Using the suitable wet etchants, the damaged material along the track can be removed more quickly than the bulk material, thus developing the tracks into nanochannels. Under the suitable conditions, channels down to a few nanometers in diameter can be produced. For example, to produce a conical nanochannel, etching was performed only from one side, the other side contained a solution that is able to neutralize the etchant as soon as the channel opens, thus slowing down the further etching process. To additionally stop the etching, the voltage used to monitor the etching process was applied in such a way that the negatively charged ions of the etchant were drawn out of the channel tip. The large opening of the conical nanochannel was called base, while the small opening was called tip. During the etching process, negatively charged carboxyl groups, which were attached to flexible polymer chains, were created on the channel surface [43].

The approaches for the preparation of ion-track-etched polymer nanochannels with different shapes have been studied and described in detail by Apel and Jiang et al. [30, 44–47]. These approaches provide a useful platform for the further development of smart nanochannel materials.

1.2.3.2 Inner-Surface Physicochemical Modifications of Nanochannels

Inspired by ion channels, in which the components are asymmetrically distributed between the membrane surfaces, the generation of biomimetic smart nanochannels is a broad and varied research field. The design and development of new biomimetic channels includes the use of different shapes of channels, different stimuli-responsive molecules, and different symmetric/asymmetric modification methods.

Herein, I suggest three routes for the design and preparation of various artificial functional nanochannels: symmetric/asymmetric design for building the shapes of the nanochannels, the physicochemical modification of the inner surface of nanochannels, and the above two methods combined in a co-design strategy (Fig. 1.7). In these three routes, the idea of asymmetric design provides more flexible approaches for building various functional nanochannels. In this chapter,

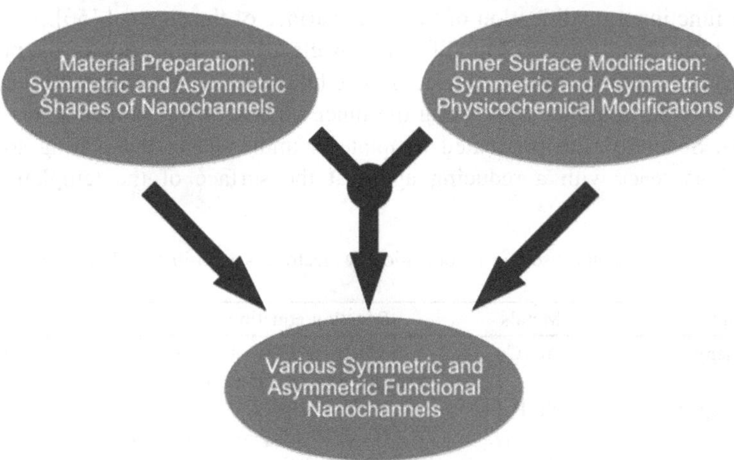

Fig. 1.7 The symmetric/asymmetric design and preparation of bio-inspired artificial functional nanochannels. Reproduced from Ref. [47] by permission of John Wiley & Sons Ltd

the asymmetric design is reflected mainly in the symmetric and asymmetric physicochemical modification. The approaches for the preparation of nanochannel materials have been described in detail by Dekker [6], Matile et al. [27], Gyurcsanyi [39], Apel et al. [44], and Martin et al. [48].

Symmetric Modification of Nanochannels

Symmetric modification of nanochannels is the most commonly used method, because it has the advantage of simple and direct implementation of modification of the entire inner surface of the nanochannels. It is universally suitable for making a variety of modifications to materials. It is also noteworthy that this modification method can take advantage of the symmetric/asymmetric shape of the nanochannel itself to achieve symmetric/asymmetric regulation of ion transport inside the nanochannel. The major symmetric modification methods are illustrated by the following examples: electroless deposition, chemical modification by functional molecules in solution and self-assembly.

Electroless Deposition

Electroless deposition is a nongalvanic type of plating method that involves several simultaneous reactions in an aqueous solution. It occurs without the use of external electrical power [49]. According to the principle of redox reactions, the use of strong reducing agents in a solution that contains metal ions results in reduction of the metal ions to metal atoms and deposition of the metal atoms onto a surface to form a dense coating. Various metals can be used for modification of nanochannels by electroless deposition, such as gold, nickel, silver, copper, palladium, and platinum. Typical chemical reaction equations for this process are shown in Table 1.1.

Based on this approach, Martin and co-workers reported a metal-polymer composite nanochannel (Fig. 1.8a), which provides a good basic platform for the further functional modification of the inner surface of the channel [50]. As shown in Fig. 1.8b, the processes of modification by electroless deposition of the polymer nanochannel are divided into two steps: The first step is pretreatment of the track-etched polymer template by coating the inner surface of the nanochannels with a catalyst. Secondly, the pretreated template is immersed in the plating solution. Metal ions react with a reducing agent at the surface of the template. After

Table 1.1 Typical metals used for modification by electroless deposition and the related reaction equations

Reductants	Metals	Reaction equations	Ref.
Formaldehyde	Au, Cu	$2Au^+ + HCHO + 3OH^- \rightarrow HCOO^- + 2\,H_2O + 2\,Au$	[50–52]
Hypophosphite	Ni, Pd	$Ni^{2+} + 2H_2PO_2^- + 2\,H_2O \rightarrow Ni + 2H_2PO_3^- + 2\,H^+ + H_2$	[53, 54]
Borohydride	Pt, Au	$2Pt^{2+} + BH_4^- + 4OH^- \rightarrow 2\,Pt + BO_2^- + 2H_2 + 2\,H_2O$	[49, 55]

Reproduced from Ref. [47] by permission of John Wiley & Sons Ltd

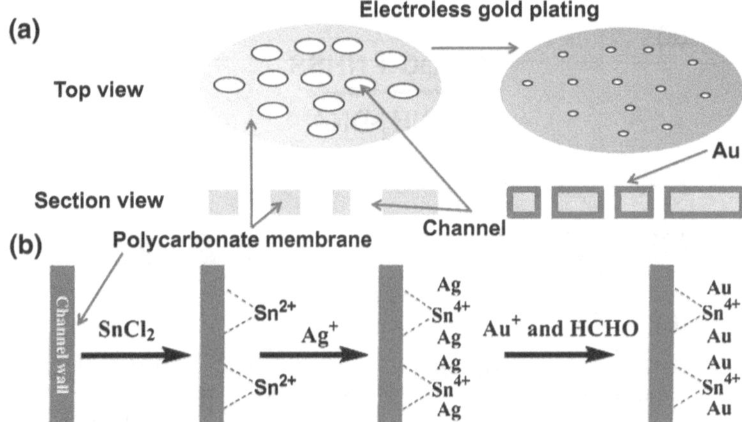

Fig. 1.8 **a** Diagram of electroless gold plating of nanochannels. **b** Chemical processes of gold electroless deposition. Reproduced from Ref. [47] by permission of John Wiley & Sons Ltd

electroless plating with the metal, the diameter of channels decreases. However, the second plating step can take a long time to achieve small enough diameters in the channels, and several toxic solvents are also used [51].

Chemical Modification by Functional Molecules in Solution

In this chapter, chemical modification mainly refers to the chemical reactions between the inner surface of the nanochannel and functional molecules in a solution to create new covalent bonds. At present, there are two main methods for chemical modification of nanochannels. The first one is based on functional groups (for example, -COOH) in the nanochannels reacting with decorating molecules to form covalent bonds (for example, -C(O)-NH-). Based on this method, Jiang and co-workers developed a pH-gating, asymmetric, single PET nanochannel by using a DNA molecular motor that was attached to the channel by a two-step chemical reaction (Fig. 1.9a) [56]. During the preparation of the PET nanochannel, negatively charged carboxyl groups were created on the inner surface of the nanochannel, which is a process that can also be performed with other similar materials, such as PI. After modification, the diameter of the nanochannel also decreases, and different sizes of decorating molecules can be used to regulate the sizes of the channels. Table 1.2 shows several decorating molecules that have been used for chemical modification of polymer nanochannels, and some typical examples are shown in Fig. 1.9b. Recently, Azzaroni and co-workers reported the design and preparation of several pH-responsive nanochannel systems by chemical modification to obtain functional nanochannels for achieving control of ionic transport [23, 57, 58].

The second method for chemical modification of nanochannels is based on self assembly of functional thiol molecules to form Au–S covalent bonds after the electroless deposition of gold onto the inner surface of the nanochannel

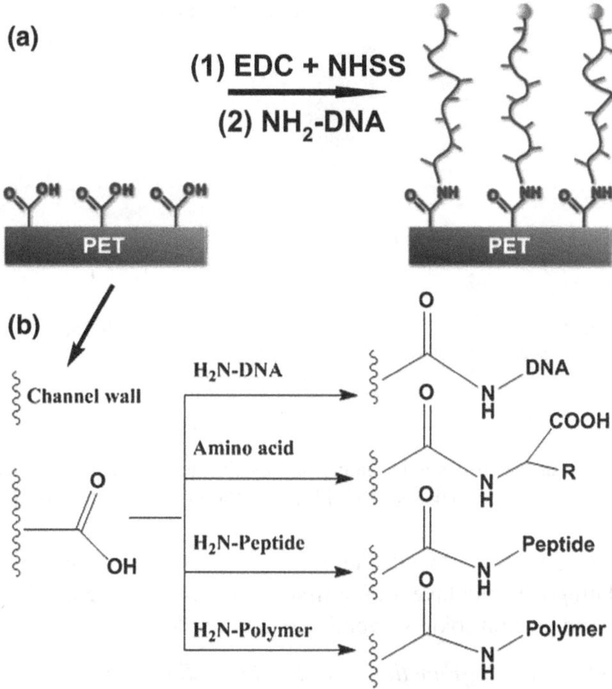

Fig. 1.9 **a** Immobilization of a DNA motor onto the inner wall of a nanochannel by chemical modification. 1-ethyl-3-[3-(dimethylamino) propyl] carbodiimide hydrochloride (EDC) and *N*-hydroxysuccinimide (NHS) are used for activating the carboxylic acid groups. **b** Several typical decorating molecules used in chemical modification of nanochannels. Reproduced from Ref. [47] by permission of John Wiley & Sons Ltd

Table 1.2 Decorating molecules used in chemical modification of polymer nanochannels by functional molecules in solution

Nanochannels	Decorating molecules	Ref.
PET	H_2N-DNA	[25, 56]
PET	L-lysine and L-histidine	[58]
PET	Poly 2-(methacryloyloxy) ethyl phosphate	[23]
PET	3-aminopropylphosphonic acid	[59]
PET	β-cyclodextrin	[60]
PI	Poly(*N*-isopropylacrylamide)	[22]
PI	Amine-PEO_3-biotin[a]	[61]
PI	Poly (methacryloyl L-lysine)	[57]
PI	Peptide nucleic acid (PNA)	[62]

Reproduced from Ref. [47] by permission of John Wiley & Sons Ltd
[a] PEO = Poly(ethylene oxide)

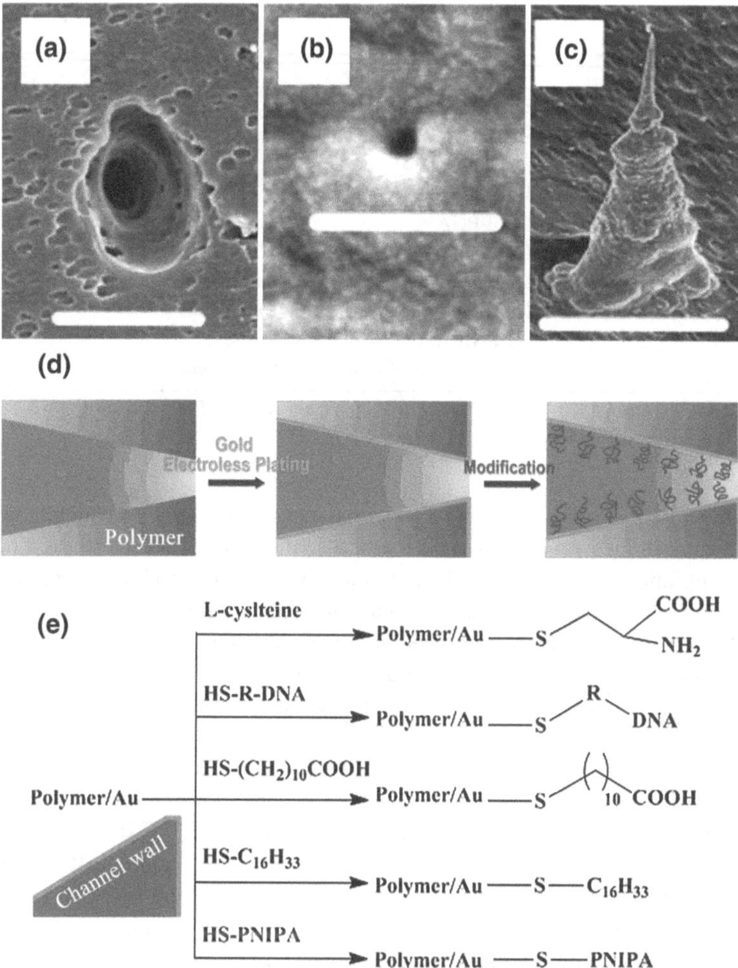

Fig. 1.10 **a-c** Electron micrographs of a single conical nanochannel after electroless modification with gold. Scale bar 5 mm. Reprinted with the permission from Ref. [64]. Copyright 2004 American Chemical Society. **d** Immobilization of functional molecules onto the inner wall of a nanochannel by electroless modification and surface chemisorbing thiols. **e** Several typical thiol self-assembled functional molecules for modifying the inner surface of a polymer/gold nanochannel. Reproduced from Ref. [47] by permission of John Wiley & Sons Ltd

(Fig. 1.10 a–d). This method has been widely used in developing many functional surfaces [63], and in the past few years has also been utilized for functionalizing nanochannels. For instance, Martin and co-workers as well as Jiang and co-workers have developed single nanochannels that are responsive to electrical potential and temperature [64, 65]. Table 1.3 shows several self-assembling thiol molecules, and some typical examples are shown in Fig. 1.10e.

Table 1.3 Several functional self-assembled thiol molecules for decorating Au-polymer composite nanochannels

Nanochannels	Decorating molecules	Ref.
PC[a]	HS-$C_{16}H_{33}$ and HS-C_2H_4-OH	[66, 67]
PC	HS-$(CH_2)_{10}$COOH	[68, 69]
PC	Cysteine	[70, 71]
PC	Thiol-DNA	[64]
PC	Thiol-PNA	[72]
PC	Thiol-antibody fragments	[73]
PC	11-mercaptoundecanoic acid and biotin	[74]
PC	Thiol- and disulfide-bearing ionophores	[75]
PET	HS-PNIPA	[65]
PET	HS-R-biotin,[b] HS-antibody, HS-Protein G	[76]

Reproduced from Ref. [47] by permission of John Wiley & Sons Ltd
[a] PC = Polycarbonate. [b] R = -$(CH_2)_2C(O)NH(CH_2)_6NH$-

Relative to the second method, the first method is direct covalent chemical modification. It has no metallization process, thus, the surface of the channels remains organic in nature, which is more closely analogous to biological channels. However, the second method is advantageous for making nanodevices, as these channels are stable with a metalized surface, and the density of the chemical modification can be better controlled as a result of the formation of a thiol/gold self-assembled monolayer. Furthermore, the second method is also slow, with the functionalization time as long as one day.

Self-Assembly

From a physicochemical point of view, modification by self-assembly is different from covalent modification with self-assembled thiol compounds. It is based on specific molecular interactions between specific groups of the inner surface of the nanochannels and functional molecules, such as electrostatic interactions [19, 77] or intermolecular forces (hydrogen bonding [62] and van der Waals forces [78]), separately or together, to achieve functionalization of the inner surface of the nanochannels [61]. However, the range of physicochemical modification that can be obtained is limited.

Azzaroni et al. reported layer-by-layer assembly of polyelectrolytes into an artificial single nanochannel that rectifies ionic current and was prepared by the electrostatic self-assembly of multiple layers of functional molecules inside the channel (Fig. 1.11a) [19]. They also reported a biosensor that contains functionalized, single, asymmetric PI nanochannels, which functions by self-recognition of ligands between a protein recognition ligand (biotin) and a specific protein molecule receptor (streptavidin, Fig. 1.11b) [61].

Asymmetric Modification of Nanochannels

Asymmetric functional modification of nanochannels is still in its early stages. In general, asymmetric modification methods are also adequate for symmetric

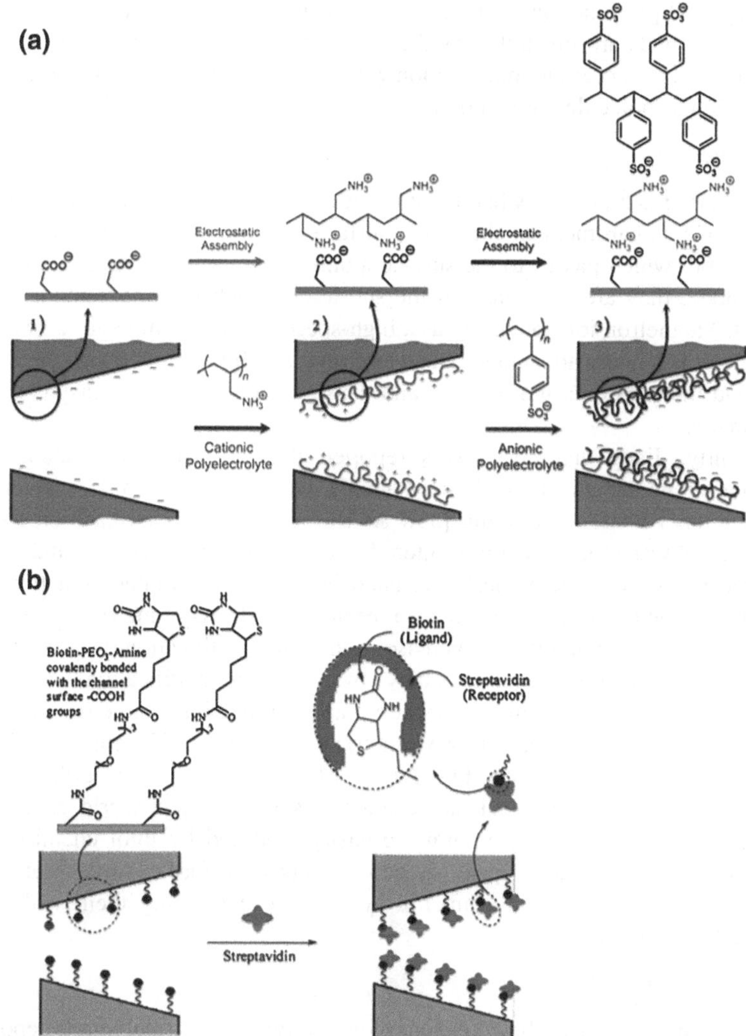

Fig. 1.11 **a** Illustration of sequential nanochannel modification through layer-by-layer assembly of polyelectrolytes: (*1*) As-synthesized nanochannel; (*2*) (Poly(allylamine hydrochloride) (PAH))$_1$(poly(styrenesulfonate) (PSS))$_0$-modified nanochannel; (*3*) (PAH)$_1$(PSS)$_1$-modified nanochannel. Reprinted with the permission from Ref. [19]. Copyright 2010 American Chemical Society. **b** Representation of an asymmetric nanochannel functionalized with biotin-PEO$_3$-amine and subsequent biospecific, noncovalent binding of a streptavidin analyte. Reproduced from Ref. [61] by permission of John Wiley & Sons Ltd

modification. However, asymmetric modification could potentially provide new ideas for developing smart nanochannel systems, because complex modification approaches can precisely functionalize diverse but specific local areas of the channels with different functional molecules. It provides more design ideas for building multiple functional nanochannel systems. The typical asymmetric modification methods are illustrated by the following examples: ion sputtering, physical/chemical evaporation, modification with plasma, and asymmetric modification by functional molecules in solution.

Ion Sputtering

Ion sputtering is a process whereby atoms are ejected from the cathode (target) as a result of bombardment of the target surface by accelerating positive ions, the energy from which passes to the surface atoms of the cathode. After atoms leave the cathode, they are deposited on the substrate, which is a momentum transfer process. Magnetron ion sputtering as a high-speed and low-temperature sputtering technology has many advantages, such as strong combination between the coating layer and the substrate, the uniform and dense coating layer, and easy control of the process.

Recently, Jiang and co-workers reported the fabrication of stable, single asymmetric nanochannels with controllable ionic rectification by asymmetric modification with ion sputtering [31]. As shown in Fig. 1.12a, each side of the nanochannel was independently sputtered to control the pore size for studying ion transport properties. The properties of nanochannels that have been functionalized by double-sided sputtering have also been studied. Asymmetric modification approaches impart a significant increase in the ionic rectification properties of the nanochannel, which is difficult to achieve by symmetric modification. On the basis of the above study, Jiang and co-workers further utilized a symmetric, hour-glass shaped single nanochannel with asymmetric sputtering of different metal ions to develop ionic rectifier systems (Fig. 1.12b), which moves one step further towards the development of functional nanochannel systems for real-world applications [79]. Moreover, gold and platinum are easily modified by thiol chemisorption, thus, these nanochannels open up new avenues to further investigate more sophisticated biochemical asymmetric systems by combining them with DNA, peptides, and proteins.

Physical/Chemical Evaporation

Modification by electron-beam evaporation is a typical physical vapor-deposition technology. It is widely used for advanced surface modification. The working principle is based on bombarding a target anode with an electron beam that is given off by a charged tungsten filament under high vacuum. The electron beam causes atoms from the target to transform into the gaseous phase. These atoms then precipitate into the solid state, which coats the surface of the substrate with a thin layer of the anode material. Characteristics of this method are fast deposition and a clean surface after modification. In particular, the coating layer is strongly adhered to the substrate, and the choice of substrate material is unlimited.

Based on this technology, Siwy and co-workers successfully prepared an ionic transistor by asymmetric electron-beam evaporation [20]. As shown in Fig. 1.13, titanium as the adhesive layers, gold as the gate electrode, and SiO_2 as the insulating layer were all deposited on the side of a conical nanochannel with a small

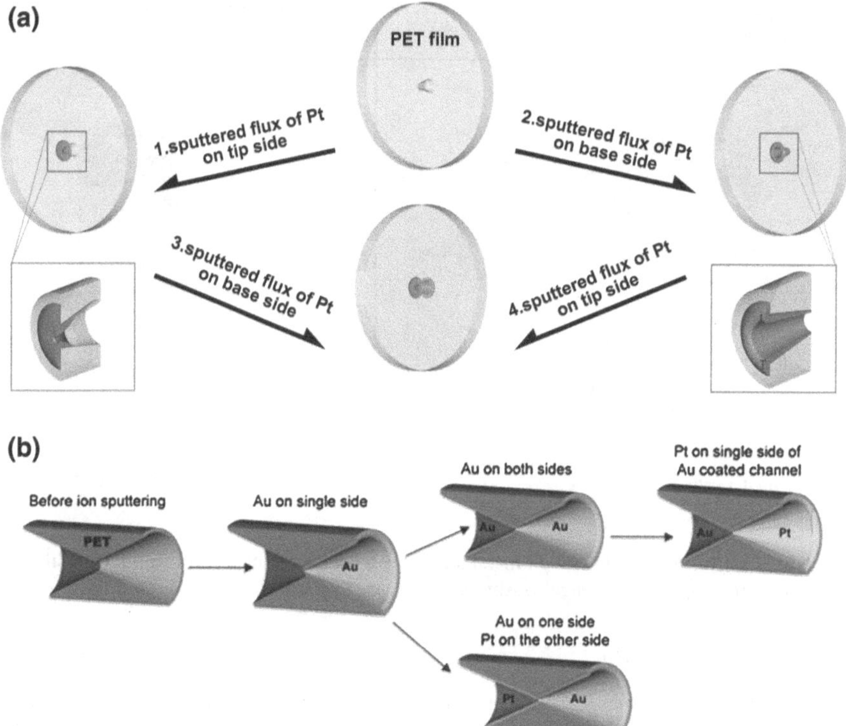

Fig. 1.12 Metal-polymer composite asymmetric nanochannels. **a** The experimental design. Reproduced from Ref. [31] by permission of John Wiley & Sons Ltd. **b** Four hour-glass shaped nanochannels composed of gold or platinum deposited on different sides of the nanochannels. Reprinted from Ref. [79], Copyright 2011, with permission from Elsevier

opening. The gold layer serves as the gate electrode, which is used to supply the chosen potential at the surface of the nanochannel. By changing the electric potential applied to the gate, the current through the channel can be changed from the rectifying behavior of a typical asymmetric conical nanochannel to the almost linear behavior of a symmetric nanochannel.

Initiated chemical vapor deposition is a chemical vapor-deposition method that directly translates free radical polymerization into a chemical vapor deposition process [80]. The process does not involve any solvents, hazardous byproducts, or other contaminants, and so it is environmentally benign. Because the reactants are delivered in the vapor phase, the modification process is not subject to surface tension limitations, and uniform, pinhole-free coatings can be formed on essentially any substrate. Based on the above method, Asatekin and Gleason successfully developed functional polymeric nanochannels for hydrophobicity-based separations [81].

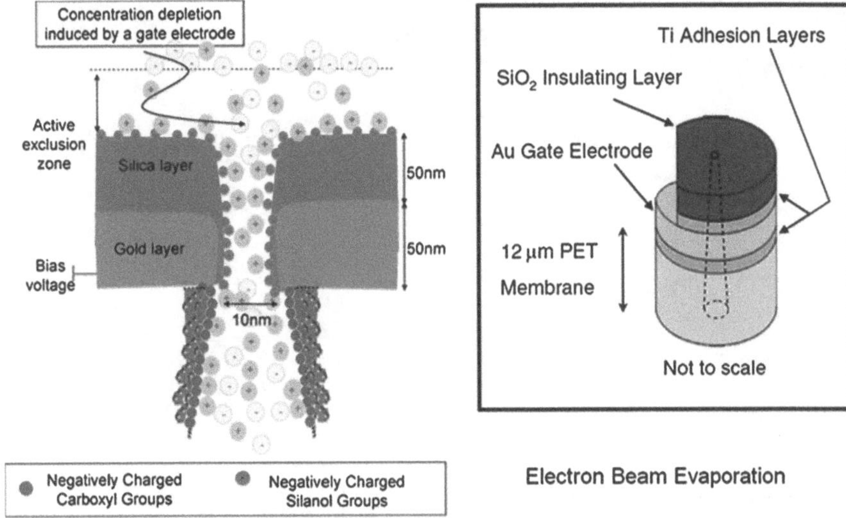

Fig. 1.13 Conical nanochannel with metal and SiO_2 layers. Reprinted from Ref. [20], with kind permission from Springer Science+Business Media

Modification with Plasma

Modification with plasma offers an effective method for nanoscale surface engineering of materials, which can functionalize a specific local area precisely, whether by symmetric or asymmetric chemical modification. This can provide a variety of nanoscale features and properties for developing advanced nanomaterials. The principle is based on plasma modification of polymer materials and can effectively produce large amounts of free radicals in the surface layer. The newly generated free radicals continue to react to form functional groups. However, the structures that are obtained by plasma modification within the nanochannels are not well characterized, and the density of functional groups is not well understood.

Jiang and co-workers reported building a pH-gating single nanochannel for ionic transport by using asymmetric chemical modification with plasma to achieve the asymmetric modification of a symmetric channel [82]. Most recently, Jiang and co-workers further utilized this method to develop a biomimetic, asymmetric, dual-responsive single nanochannel by using asymmetric plasma modification approaches to functionalize two sides of the channel with different functional molecules. This method provides simultaneous control over the pH- and temperature-dependent asymmetric ion-transport properties inside the channel (Fig. 1.14) [83].

Asymmetric Chemical Modification by Functional Molecules in Solution

Asymmetric chemical modification by functional molecules in solution is based on the minimum concentration requirements of the chemical modification reaction, and the different reagent concentration distribution inside the symmetric/asymmetric confined space of the nanochannels. Therefore, the different shapes of the

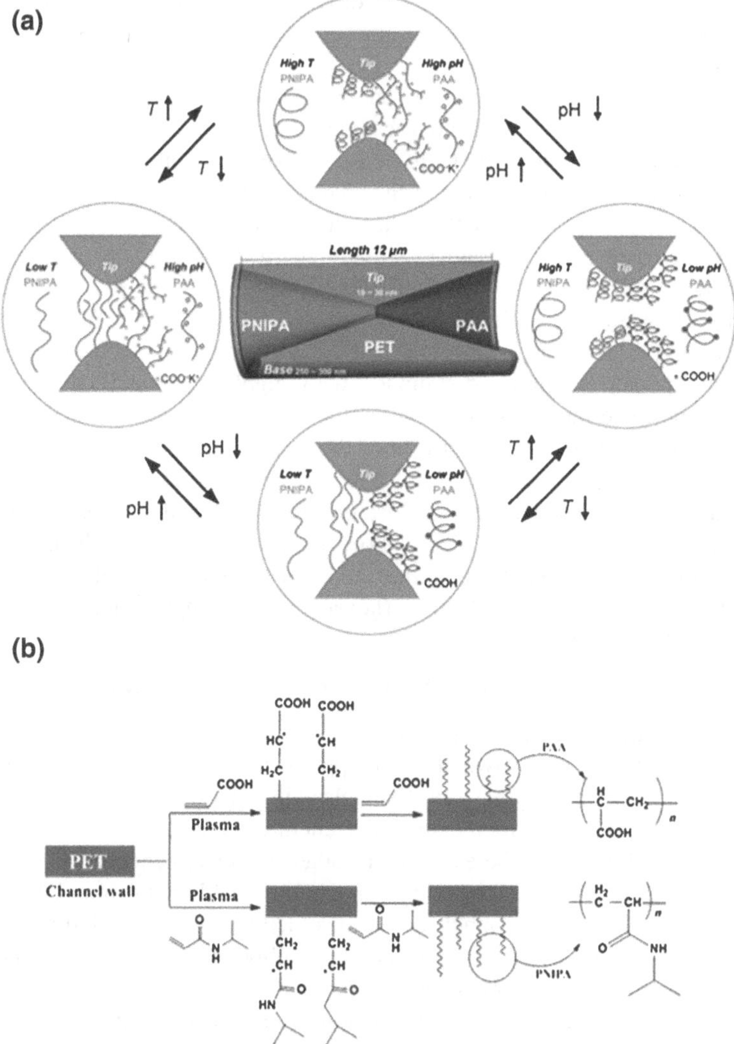

Fig. 1.14 **a** A biomimetic asymmetric pH/Temperature responsive single nanochannel produced by asymmetric plasma modification. One side of the nanochannel was treated by plasma-induced grafting of *N*-isopropylacrylamide in the vapor phase, which became a temperature-responsive polymer, PNIPA, after plasma-induced graft polymerization. The other side of the nanochannel was treated by plasma-induced grafting of distilled acrylic acid in the vapor phase, which became a pH-responsive polymer, polyacrylic acid (PAA), after plasma-induced graft polymerization. Reprinted with the permission from Ref. [83]. Copyright 2010 American Chemical Society. **b** Plasma polymerization modification processes of the inner surface of the nanochannel. Reproduced from Ref. [47] by permission of John Wiley & Sons Ltd

nanochannels can be used to achieve a different concentration distribution and, consequently, symmetric and asymmetric chemical modifications of the nanochannels.

Vlassiouk and Siwy developed the above method for asymmetric modification of nanochannels to build a nanofluidic diode that relies on a concentration gradient. The concentration gradient is constant over time and is created when a reagent is placed only on one side of the channel [84]. Figure 1.15 shows a single conical nanochannel in which the inner surface was patterned so that a sharp boundary between positively and negatively charged regions was created. Later, Siwy and co-workers further utilized this strategy to develop a bipolar ionic transistor by using a single hour-glass shaped nanochannel with asymmetric modification [38]. Recently, Siwy and co-workers also reported a biosensor that is based on the above nanofluidic diodes with highly nonlinear current/voltage characteristics, and the nanodevices allow the isoelectric point of minute amounts of proteins immobilized on the surface of the nanochannel to be determined [85].

A solution assembly method is also a possible way to achieve asymmetric modification inside nanochannels. Most recently, Xue and co-workers reported that the degree of ionic current rectification of a nanochannel could be finely tuned over a wide range by controlling the modified region of the channel and the concentration of cationic surfactant (hexadecyl trimethylammonium bromide) inside the channel [86].

Selective Strategies for Various Modifications of Nanochannels

In order to further demonstrate how to choose a suitable modification method to functionalize nanochannels, two classical approaches for chemical modification of the interior surfaces of polymer nanochannels with functional molecules are shown in Fig. 1.16 as examples to illustrate the chemical modification with selective strategies, which may also be extended to other nanopore/nanochannel materials.

The first approach is that functional molecules are directly immobilized onto the inner surfaces of nanochannels by various methods, such as solution chemical covalent modification, electrostatic self-assembly and plasma modification. Compared with the second approach, characteristics of this approach are that it is more convenient to achieve a direct function of channels, and these channels are more closely analogous to biological channels [87].

The second approach consists of two steps: First step, metallization of the channel surface using electroless deposition, ion sputtering deposition, or electron beam evaporator. Second step, covalent modification through forming self-assembled monolayers from functional molecules, due to the spontaneous formation of S–Au or S–Pt bonds between molecules possessing SH or S–S groups and the metal surface [88–90]. It is generally believed that this approach is advantageous for making nanodevices, as it is a good way to solve the first approach concerning the density of the chemical modification which is hard to control [13, 39], and also these channels are very stable because of the metallized surface [31].

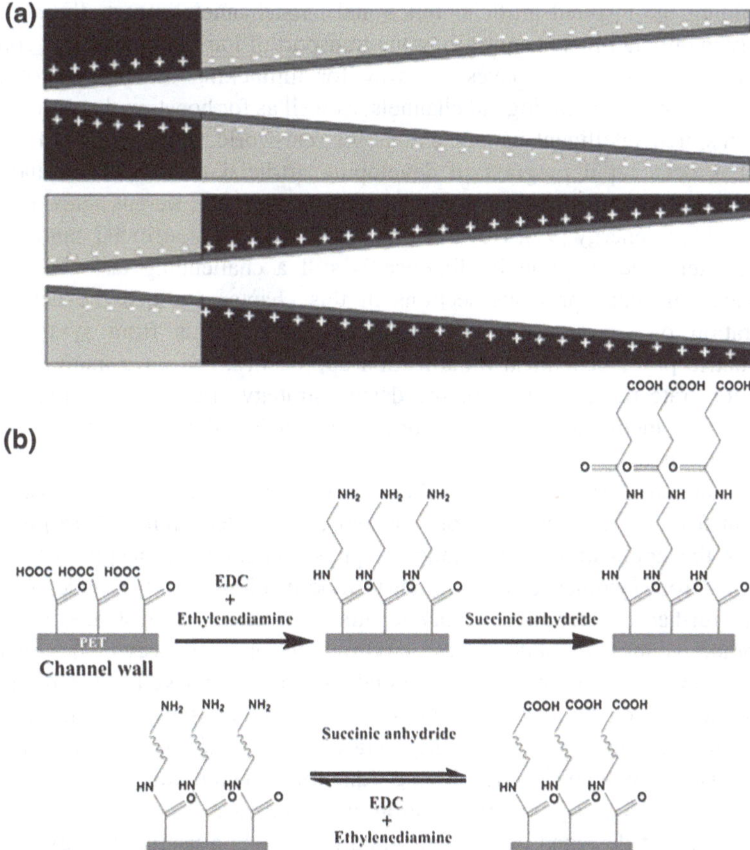

Fig. 1.15 **a** Asymmetric chemical modification by functional molecules in solution as applied to transform carboxylic acid groups into amino groups with different concentration gradients onto the inner wall of the conical nanochannels. Reprinted with the permission from Ref. [84]. Copyright 2007 American Chemical Society. **b** Chemical modification of the inner surface of the nanochannel by functional molecules in solution. Carboxylic acid groups can be transformed into amide groups with EDC and ethylenediamine, and the resulting surface amines can be transformed to carboxylic acid groups by succinic anhydride. Reproduced from Ref. [47] by permission of John Wiley & Sons Ltd

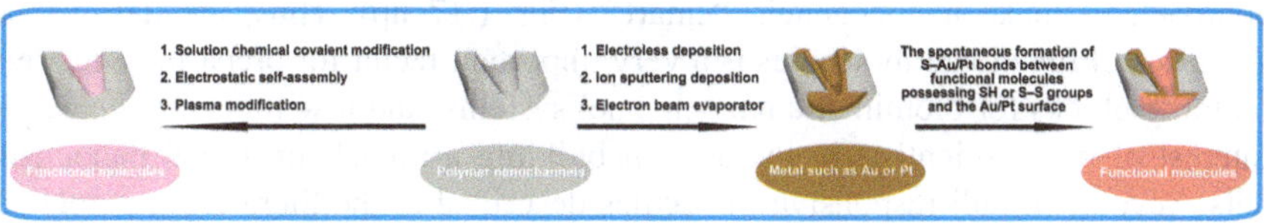

Fig. 1.16 Two classical approaches used in chemical modification of the interior surfaces of polymer nanochannels with functional molecules. Reproduced from Ref. [9] by permission of The Royal Society of Chemistry

Building bio-inspired artificial functional nanochannels imparts the ability of actively manipulating and controlling the transport of ions and molecules confined in a nanoscale space. It paves the way for mimicking the process of ionic/ molecular transport in biological channels, as well as for boosting the development of bio-inspired intelligent nanomachines for real-world applications. At present, there has been rapid progress in developing artificial nanochannels, and these nanochannels are transforming technology in a variety of different fields, from life science [14] to energy [91]. However, how to endow these artificial nanochannels with greater functions and intelligence is still a challenging task. Inspired by biological channels, previous sections in this chapter outlined the design and preparation of various artificial functional nanochannels from symmetric to asymmetric physicochemical modification approaches, and gave many examples to demonstrate the feasibility of the design strategy. Table 1.4 summarizes the above typical modification methods for nanochannels and gives comments on each of these methods.

The ion transport properties of these artificial nanochannels are substantially different from those of macroscopic channels, and understanding these properties requires the application of fundamental physicochemical concepts. The design strategies from symmetric to asymmetric modifications could provide platforms for the further development of nanochannel systems. For instance, the transmembrane asymmetric states of the environment are omnipresent in cell membranes. However, at present, most artificial nanochannels have been studied under symmetric stimuli, and how to achieve various smart transport controls under asymmetric stimuli is still in its early stages. Asymmetric modification will provide a useful method for developing functional nanofluidic devices that can operate in complex, asymmetric solution environments.

I expect that designed asymmetric modification as a useful strategy will find broad applications in the development of artificial functional nanochannel systems and facilitate the asymmetric design of other materials.

1.3 Typical Examples of Responsive Single Nanopore/Nanochannel

The physicochemical modification with functional molecules is a primary approach to make nanochannels "smart" (Fig. 1.17 up). Thus, the design and synthesis of functional molecules is a very important factor for preparing excellent building blocks for biomimetic nanochannel systems, and it will inspire the active interest of many scientists in the area of building artificial functional nanochannels. Several stimuli-responsive molecules described in the literature, i.e., pH [2, 21, 23, 57], temperature [2], light [2, 15], ions [94, 95] are listed in Fig. 1.17 (bottom).

Table 1.4 Comments on various typical modification methods for nanochannels

Modification method	Nanochannels	Functional substances	Comments	Ref.
Electroless deposition	PC, PET	Au	Increases the stability of nanochannels by plating a metal layer on the inner surface of the nanochannels and provides a good platform for further binding of all kinds of functional molecules by the self-assembly of thiols and chemical modification. However, this method is time-consuming and involves toxic chemicals and heavy-metal salts	[50, 92]
Chemical modification by functional molecules in solution	PET, PI, PC, PC(PET)/Au	DNA, biotin, amino acids, polymeric brushes, thiols	The most commonly used method for functionalizing the inner surface of nanochannels through stable covalent bonds, such as C(O)-NH bonds and Au–S interactions	[21, 23, 56–58, 62, 70, 71, 76, 93]
Self-assembly	PET, PI	Polyelectrolytes	Changes the inner surface of nanochannels by specific molecular interactions, such as electrostatic interactions and intermolecular forces. These functional nanochannels are greatly influenced by environmental factors because of noncovalent interactions. The range of functionalities that can be obtained is limited	[19]
Ion sputtering	PET	Au, Pt	Coats the inner surface of nanochannels by depositing metal layers, increases the stability of the nanochannels, and provides a platform for further binding of all kinds of functional molecules by the self-assembly of thiols. It can be easily and widely used to achieve asymmetric modification of nanochannels. However, the thickness and uniformity of deposited metal layer within the channels is not well understood	[31, 79]
Electron-beam evaporation	PET	Au, SiO_2	Coats the surface of nanochannels by depositing metal/inorganic layers and can easily get continuous, multiple, nanometer-thick layers of metals	[20]

(continued)

Table 1.4 (continued)

Modification method	Nanochannels	Functional substances	Comments	Ref.
Initiated chemical vapor deposition	PC	Polymers	Coats the inner surface of nanochannels by depositing nanometer-thick polymer layers, which can easily and quickly be used to symmetrically/asymmetrically modify any substrate without surface-tension limitations. It is also environmentally benign. However, the stability of the deposited polymer layer is not good	[81]
Plasma	PET	Polymer brushes	Controls specific modification of areas of nanochannels by symmetric and asymmetric plasma grafting of various polymers. It is easy and fast to obtain functional nanochannels; however, the modification density is hard to control	[82, 83]
Asymmetric chemical modification by functional molecules in solution	PET	Antibodies, succinic anhydride, ethylenediamine	Commonly used method for functionalizing diverse but specific local areas of the nanochannels through stable covalent bonds. The shape and the surface charge of the original nanochannels have a great impact on the asymmetric modification of the channels	[38, 84, 85]
Asymmetric assembly in solution	PET	Surfactants, particles	It is a possible solution method for functionalizing diverse but specific local areas of the nanochannels by physical adsorption of surfactants or particles. This method is cheap, versatile and the materials are reusable	[86]

Reproduced from Ref. [47] by permission of John Wiley & Sons Ltd

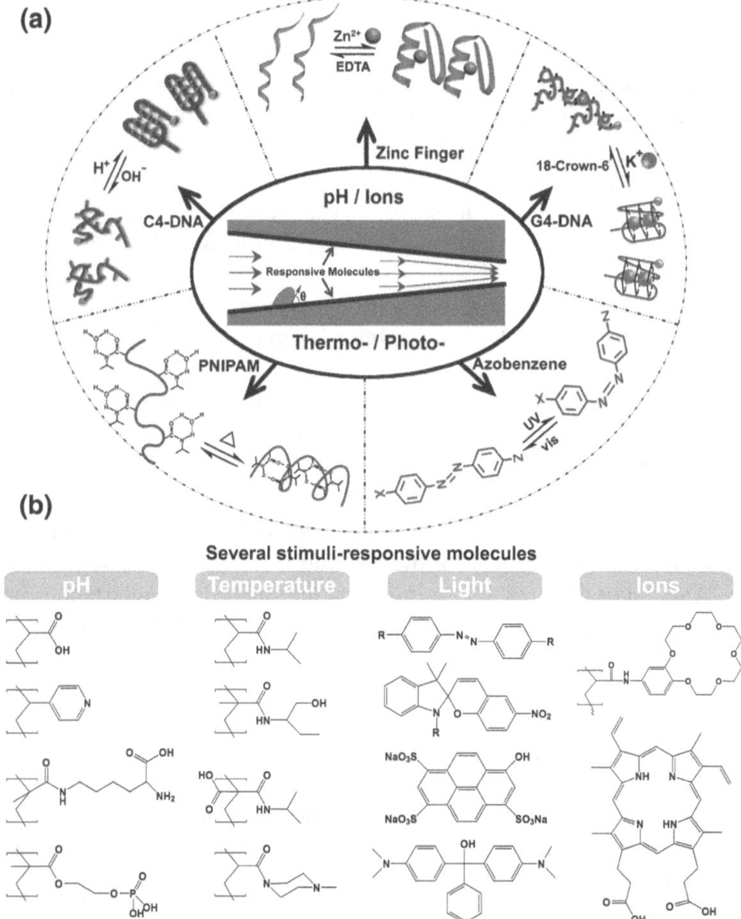

Fig. 1.17 Scheme of smart gating of nanochannels with functional molecules which can respond to the external stimuli, e.g., pH, temperature, light and ions. Reproduced from Ref. [15] by permission of John Wiley & Sons Ltd. Reproduced from Ref. [9] by permission of The Royal Society of Chemistry

In the next section, I will provide a comprehensive overview of simple and multiple responsive single nanopore/nanochannel materials in this rapidly growing field, from biological, inorganic, organic to composite nanopore and nanochannel materials, which can respond to single/multiple external stimuli, e.g., pH, temperature, light, ions and so on.

1.3.1 Basic Concepts and Theories

Ionic transport properties of ion channels play a crucial role in many important physiological processes in cells, which are mainly composed of three characteristic features including ionic selectivity, ionic rectification and ionic gating. These characteristics have attracted broad interest from scientists in various research fields. Similarly, the interest in artificial nanochannel systems stems from the crucial role of ionic transport phenomena through biological channels, and has also attracted wide attention [96].

Ionic selectivity is defined to evaluate the performance of a nanochannel for ionic flow control [88, 97]. Ionic selectivity is the ratio of the difference in ionic currents of majority and minority carriers to the total ionic current carried by both positive and negative ions. For a nanochannel with perfect control over cation and anion, the selectivity is uniform. For a nanochannel without ionic flow control, the selectivity is zero.

The effect of ionic rectification is observed as asymmetric current–voltage (*I–V*) curves (Fig. 1.18), with the ionic current recorded for one voltage polarity higher than the current recorded for the same absolute value of voltage of opposite polarity. This diode-like *I–V* curve indicates that there is a preferential direction for ionic flow [96]. Several mechanisms have been described to explain the ionic rectification effect observed in artificial nanochannel systems, such as the electrostatic model based on the concepts of a rocking ratchet and an electrostatic trap [96, 98, 99], Woermann mode [100, 101], the quantitative description of ionic current through a nanochannel by the Poisson and Nernst–Planck equations [102–106], and the electrostatic model based on polyelectrolyte theory [107].

Ionic gating is defined to evaluate the performance of ion passing through the nanochannel that is governed by a "gate", which may be opened or closed in response to chemical or electrical signals, temperature, or mechanical force [56, 82].

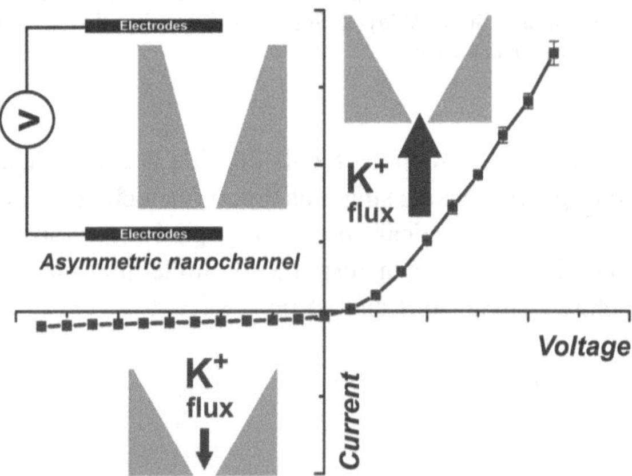

Fig. 1.18 Ionic rectification in a nanochannel with broken symmetry

1.3.2 Simple Responsive Single Nanopore/Nanochannel

There has been rapid progress in developing nanopore/nanochannel materials that respond to a simple external stimulus, such as pH, temperature, light, electric potential, ions and molecules. In the following sections, each of the six external stimuli of regulating ionic transport properties is illustrated with different types of materials in the order of biological materials, organic materials and composite materials.

1.3.2.1 pH Responsive Single Nanopore/Nanochannel

At present, pH responsive nanopores/nanochannels have been the most extensively studied materials for two main reasons as follows: First, pH is a very significant value for all electrochemical reactions, and the change of pH is recognized as one of the most important biochemical changes for ion channels in life processes. Second, pH change can be easily achieved by a simple acid and alkali regulation approach.

Aguilella's group reported the construction of a pH responsive biological nanopore by using a lipid membrane reconstituted protein pore, and the bacterial porin OmpF inside the external membrane of *Escherichia Coli* (Fig. 1.19a, inset) [89]. This nanopore exhibits linear *I–V* curves (Fig. 1.19a) under symmetric pH conditions, and ionic transport properties of the pore are modified during the symmetric change of pH from 3 to 12, which means that the pore has a certain ionic gating property with different pH. Meanwhile, there was a remarkable difference in that significant rectifications were observed as *I–V* curves under asymmetric pH conditions (Fig. 1.19b). The pH responsive feature of the pore lies

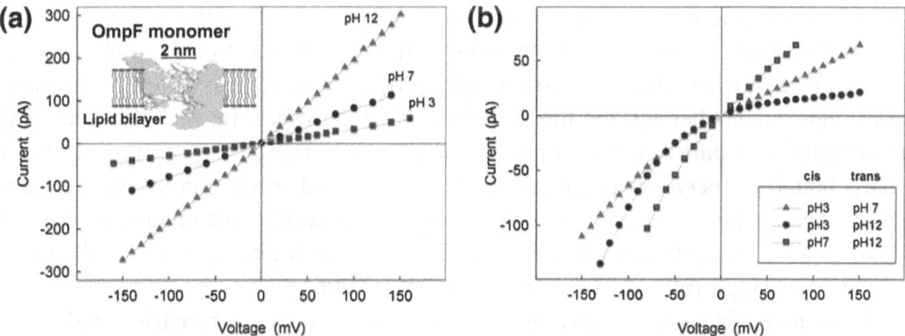

Fig. 1.19 Rectifying properties of a pH-tunable single biological nanopore. **a** *I–V* curves of OmpF pore in 0.1 M KCl under symmetric pH conditions. Labels indicate the pH on both solutions from pH 3 to pH 12. The inset shows axial section of the OmpF pore in a lipid bilayer. **b** *I–V* curves of OmpF pore in 0.1 M KCl under asymmetric pH conditions show rectified ionic current behavior. Labels indicate the pH on the cis side and the trans side. Reprinted with the permission from Ref. [89]. Copyright 2006 American Chemical Society

on the particularities of the reversible protonation/deprotonation of the reconstituted protein pore titratable residues. This pH responsive feature enables the cation/anion selection between linear and rectifying ionic currents by changing pH under symmetric or asymmetric conditions. However, this biological nanopore involves a serious limitation for technological applications due to the instability of the bilayer lipid membranes.

Compared with the biological materials, organic materials, such as polymers, have greater flexibility in terms of shape and size, superior robustness, and surface properties. When the size of the pore or channel is small within a nanoscale range, there are two key factors to control ionic transport properties inside the nanopore/nanochannel: the shape and the inner surface chemical properties of the pore/channel [8, 39].

According to the first factor, asymmetric shaped single nanochannels have been prepared in PET by applying the asymmetric track-etching technique [43]. During the etching process, negatively charged carboxyl groups, which were attached to flexible polymer chains, were created on the channel inner surface. When the pH of electrolyte is higher than the isoelectric point of the etched polymer surface, the surface will be negatively charged [43]. For lower pH values the surface is protonizing of the groups and adsorption of protons, and it shows the loss of the rectification properties [87]. As shown in Fig. 1.20a, the conically shaped single nanochannel in PET behaves as asymmetric *I–V* characteristic where the preferential direction of the cation flow is from the narrow entrance towards the wide aperture of the channel under neutral and basic conditions. The ionic rectification phenomenon is pH sensitive. Many other materials also have similar properties, such as polyimide (PI) [43] and glass pipettes [108].

There are many models to explain the ionic rectification of these nanopore/nanochannel systems e.g., models for ionic rectification in nanopore/nanochannel with diameters comparable to the thickness of the electrical double layer. It is generally thought that the surface charge, that is, the electrostatic interaction, plays an important role in regulating ionic transport properties of the nanopore/nanochannel systems. Siwy has already reviewed ionic rectification in single nanopores/nanochannels with asymmetric shapes [96]. Her group also developed a single asymmetric nanochannel, the inner surface of which was patterned after chemical modification to partially transform carboxyl groups into amino groups so that a sharp boundary between positively and negatively charged regions was created [84]. The rectification property is not a typical pH responsive change of the original conically shaped nanochannel, and there is a maximum value in middle range of the pH change from acidic to alkaline (Fig. 1.20b).

Jiang and Wang et al. developed a pH gating of an asymmetric single PET nanochannel using the pH-sensitive DNA molecular motor, which was attached into the channel [56]. It is known that the DNA molecules are negatively charged in solution, similar to the carboxyl groups in the original PET nanochannel. However, there is another important factor that contribute to the remarkably enlarged difference in ionic current at low and at high pH. In low-pH solution, the motor DNA folds into a densely packed rigid quadruplex i-motif structure that

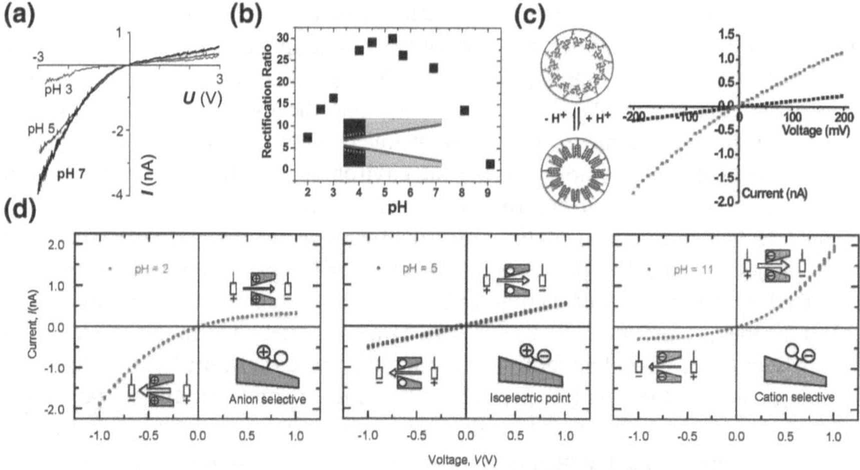

Fig. 1.20 Ionic transport properties of pH responsive polymer asymmetric nanochannels. **a** *I–V* characteristics of the original PET single nanochannel recorded at symmetric electrolyte conditions. The measurements were performed in 0.1 M KCl at different pH. Reprinted from Ref. [87], Copyright 2003, with permission from Elsevier. **b** Partial interior surface of the PET single nanochannel after chemical modification applied to transform carboxyl groups into amino groups. Rectification ratio at +2/−2 V with different pH of 0.1 M KCl solutions. The inset shows pattern of surface charge of the nanochannel. Reprinted with the permission from Ref. [84]. Copyright 2007 American Chemical Society. **c** *I–V* properties of the PET single nanochannel after attachment of motor DNA molecules onto the inner channel wall in phosphate buffered saline (PBS) solution. *I–V* characteristics were recorded under symmetric electrolyte conditions at pH 4.5 (*black*) and 8.5 (*red*). Cartoon describing DNA conformation is responsive to pH. Reprinted with the permission from Ref. [8], Copyright 2008 American Chemical Society. **d** *I–V* curves of the amphoteric (lysine) nanochannel at pH 2 (*left*), pH 5 (*center*), and pH 11 (*right*). Reprinted with the permission from Ref. [10]. Copyright 2009 American Chemical Society

partially decreases the effective diameter of the nanochannel (Fig. 1.20c, down). In high-pH solution, the motor DNA relaxes to a loosely packed single-stranded and more negatively charged structure that enhances the total ion conductivity inside the nanochannel (Fig. 1.20c, up). The structural transition of DNA molecular might well induce a change of effective diameter of the nanochannel. The relatively dense packing of DNA molecules with i-motif structure on the inner surface of the channel may efficiently decrease the effective diameter, resulting in low conductivity, while the DNA with single-stranded structure loosely packing on the channel wall cannot efficiently reduce the effective diameter, leading to high conductivity. Meanwhile, the extended single-stranded DNA conformation present at pH 8.5, where the backbone of the DNA molecule was highly charged, can help to increase the ionic conductivity near or inside the nanochannel tip, similar to the function of the surface charge. Thus, this difference in ionic conductivity at different pH values is remarkably enlarged as shown in Fig. 1.20c. These results coincided with later studies on the four-stranded G-quadruplex DNA structure of the biomimetic potassium responsive nanochannel [25].

Azzaroni and Ali et al. reported several pH-tunable asymmetric single nanochannels by the integration of polymer brushes or amphoteric molecules into the inner surfaces of the channels [23, 57, 58]. The growth of zwitterionic polymer brushes provides a useful approach to finely tune the ionic rectification characteristics of the nanochannels. They demonstrated that manipulating ion transport through the channel by simply varying the environmental pH is achievable, and it enables a higher degree of control over the ionic transport properties inside the channel [57]. They described the construction of a pH responsive asymmetric single nanochannnel, the permselectivity and rectifying properties of which are enhanced by using polyprotic polymer brushes as highly tuneable building blocks [23]. L-lysine as amphoteric chains was also immobilized onto the inner surface of the single conical nanochannel to finely tune the ionic transport by presetting the environmental pH [58]. At low pH values, the ionized amino groups are positively charged due to protonation, while the protonated carboxylic groups are neutral (Fig. 1.20d, left). The nanochannel is then selective to anions and shows the rectification properties of conical channel with positive fixed charges. Conversely, at high pH values, the deprotonated amino groups are in neutral form, while the carboxylic groups are ionized (Fig. 1.20d, right). The transition between these two rectification regimes occurs at pH value close to the isoelectric point. At this pH value, the carboxylic and the amino groups are charged, but the nanochannel net charge is zero, and the *I–V* curve shows a linear behavior (Fig. 1.20d, center).

In addition to the above asymmetric shaped single nanochannels, several pH responsive symmetric shaped single nanochannels have also been developed [21, 38, 82]. A single cylindrical polymer nanochannel modified with pH responsive polymer brushes was developed to achieve the ionic current gating characteristics, and its transmembrane ionic current can be accurately controlled by manipulating the proton concentration in the surrounding environment [21, 107]. The created pH switchable and tunable single nanochannels display ionic transport properties similar to the typical behaviors observed in many biological channels. These behaviors realize key pH-dependent transport functions in living organisms, where the nanochannel can be switched from an "off" state to an "on" state in response to a pH drop. Figure 1.21a depicts the *I–V* curves of this single nanochannel modified with polyvinylpyridine (PVP) brushes at different pH. By increasing the pH from 2 to 4, and finally to 10, a significant decrease in the transmembrane ionic current was observed under the same applied bias. The "on/off" switching was based on the manipulation of the surface charges of the channel walls via the protonation of the brush layer, which in turn control the channel conductivity.

A symmetric single hour-glass shaped polymer nanochannel also provides a good template for implementing ''smart'' nanochannel materials [82]. Siwy's group reported an hour-glass shaped nanochannel, inner surface of which was patterned after chemical modification to partially transform carboxyl groups into amino groups. Because the center section of the channel is out of the optimal range for chemical modification, it greatly reduces the rate at which the center section of the channel was modified [38]. The ionic current transistor properties in this channel were controlled chemically by changing the surface charge on the channel

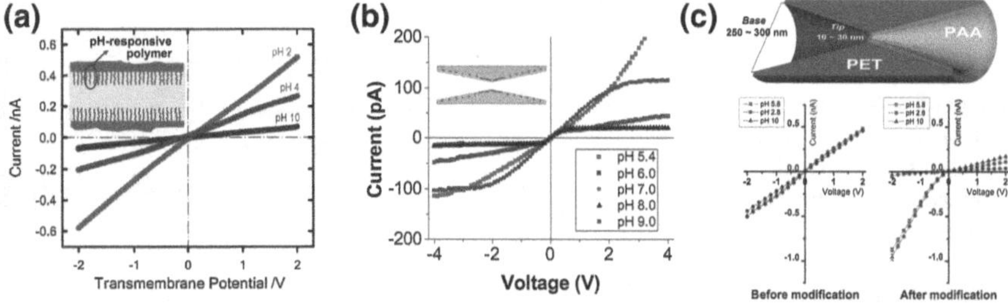

Fig. 1.21 Ionic transport properties of pH responsive polymer symmetric nanochannels. **a** *I–V* characteristics of a single cylindrical PVP brush-modified nanochannel in 0.1 M KCl at different pH. Reprinted with the permission from Ref. [21]. Copyright 2009 American Chemical Society. **b** Dependence of *I–V* curves of the nanochannel on different pH of 0.1 M KCl solution. Reproduced from Ref. [38] by permission of John Wiley & Sons Ltd. **c** *I–V* properties of the single nanochannel before and after plasma asymmetric chemical modification under different pH conditions. Reproduced from Ref. [82] by permission of John Wiley & Sons Ltd

wall. The surface charge of the nanochannel can be directly influenced by different pH. At pH 6–8, both carboxyl and amine covered parts of the channel wall were charged, forming "+–+" state, and the transistor characteristics were very distinct (Fig. 1.21b). By lowering the pH, the carboxyl groups in the center of the channel were protonated, thus creating a channel with positively charged walls near the large exterior openings and neutral surface at the center of the channel: "+0+". Raising the pH to 9 had the opposite effect; the walls near the outside of the channel became charge neutral, while the center section remained negatively charged: "0–0". It indicates that the channel with reduced charge in the center of the channel operates predominantly in the open state (*I–V* curve for pH 5.4 in Fig. 1.21b).

As mentioned above, the pH-tunable asymmetric ionic transport properties of the nanochannels [23, 57, 58] have already been successfully developed based on the asymmetric shaped nanochannel modified with pH responsive molecular architectures, and the pH-gating ionic transport properties of the nanochannels [21] were also obtained. Jiang's group further developed the pH responsive nanochannel using a plasma asymmetric chemical modification approach to achieve the asymmetric modification of the symmetric shaped single nanochannel [82]. This nanochannel is not subject to the restriction of solution environment on the chemical modification. Compared with previous materials, this responsive nanochannel has the advantage of providing simultaneous control over the pH-tunable asymmetric and pH-gating ionic transport properties. Figure 1.21c shows *I–V* properties of the nanochannel before modification, and the nanochannel exhibits linear *I–V* curves at different pH values. No change could be observed when the pH changed from 5.8 to 2.8 and from 2.8 to 10, which meant that the original nanochannel did not rectify at different pH values and did not have gating property. After asymmetric modification, there was a remarkable difference in significant rectifications, which were observed as *I–V* curves (Fig. 1.21c). By changing the pH from 5.8 to 2.8, a significant decrease in the transmembrane ionic current

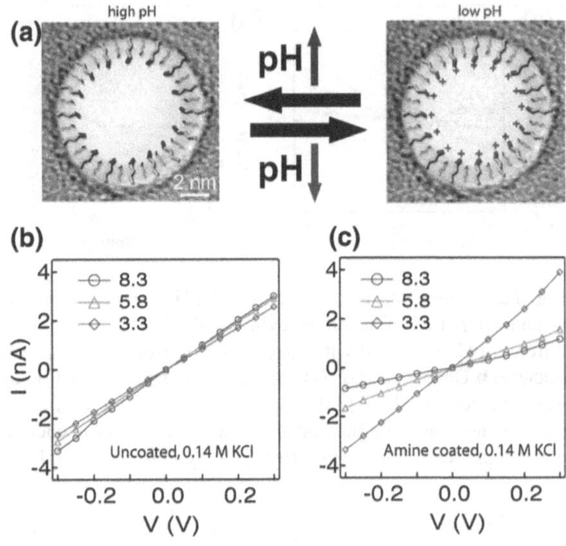

Fig. 1.22 Schematic representation of pH responsive polymer-SiN nanopore. **a** *I*–*V* curves at 0.14 M KCl for the same uncoated **b** and coated, **c** nanopore under different pH conditions. Reprinted with the permission from Ref. [109]. Copyright 2007 American Chemical Society

was observed at the same ion concentration. Then, a significant increase in the transmembrane ionic current could be observed when the pH changed from 2.8 to 10. This asymmetric ionic transport property is caused by asymmetric chemical modification, which results in structural asymmetry and non-uniform chemical composition of the nanochannel.

Once artificial inorganic nanopores/nanochannels are prepared, it is difficult to directly control ionic transport properties inside the pores or channels, due to the physicochemical properties of inorganic materials. Thus, the use of functional molecules, which were immobilized onto the inner surface of the inorganic nanopores/nanochannels as the composite materials to manipulate the ionic transport through the pores or channels, provides an attractive strategy for creating smart nanopores/nanochannels with environmental responsive characteristics. Meller's group reported a pH responsive polymer-SiN nanopore (Fig. 1.22a) by using self-assembly of organosilane molecules on the inner surface of the SiN nanopore wall [109]. *I*–*V* curves of an uncoated nanopore (Fig. 1.22b) and an amine-coated nanopore (Fig. 1.22c) were examined at 0.14 M KCl and at pH 3.3, 5.8, and 8.3. At 0.14 M KCl, the coated pore displays a marked ionic current enhancement upon lowering the pH from 8.3 to 3.3, while the uncoated pore still remains insensitive to pH. The result is in agreement with measurements performed in the track-etched original PET nanochannel, which have native carboxylic groups on its surface [43].

1.3.2.2 Temperature Responsive Single Nanopore/Nanochannel

Ion channels activated by temperature changes transduce this information into molecular conformational changes that open and close the channels. The thermosensation is carried out by the direct activation of thermally gated ion channels in the surface membranes of sensory neurons [110], and this physiological process is achieved by temperature responsive ion channels, which are members of the extensive TRP family [111]. Inspired by these ion channels, Movileanu's group developed a temperature responsive protein pore containing elastin-like polypeptide (ELP) loops [112]. These ELP loops were placed within the cavity of the lumen of the α-hemolysin pore. Below the transition temperature (T_t) the ELP loop was fully expanded and blocked the pore. Above its T_t the ELP was dehydrated and the structure collapsed, enabling a substantial flow of ions through the pore (Fig. 1.23a). The open-state current amplitude and individual current blockades of the pore were highly temperature-sensitive, and the open-state current amplitude of the ELP loop-containing RHL pore increased with the temperature from 20 to 60 °C (Fig. 1.23b, c).

Azzaroni's group reported a conically shaped single PI nanochannel modified with thermoresponsive brushes (poly(*N*-isopropylacrylamide), PNIPA) as molecular gates nanoactuated by temperature-driven conformational transitions [22]. PNIPA brushes neutralize the surface charge of the pores, resulting in the loss of the rectifying behavior and, consequently, the pore exhibits a linear *I*–

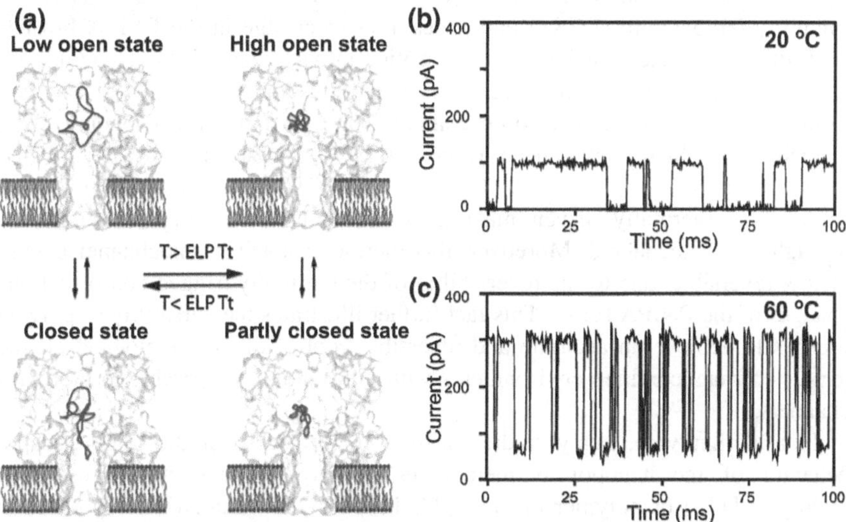

Fig. 1.23 Temperature responsive protein nanopore. **a** Model for the temperature dependent transient current blockades of the protein pore in lipid bilayers. Temperature dependence of the current through the temperature responsive pore recorded at **b** 20 °C, **c** 60 °C. The transmembrane potential was +80 mV. Reprinted with the permission from Ref. [112]. Copyright 2006 American Chemical Society

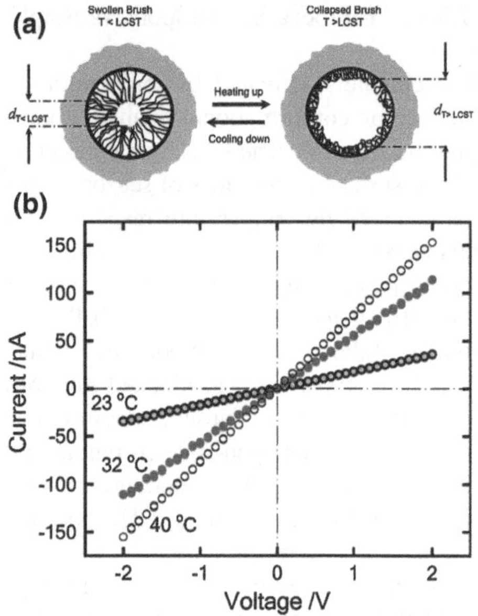

Fig. 1.24 **a** Cartoon describing the thermally driven nanoactuation of the PNIPA brushes in the nanochannel. **b** *I–V* curves in 1 M KCl for a PI conical nanochannel after modification with PNIPA brushes at different temperatures. Reproduced from Ref. [22] by permission of John Wiley & Sons Ltd

V characteristic (Fig. 1.24b). At room temperature (23 °C) PNIPA brushes remain swollen, thus decreasing the effective cross section of the nanochannel. This is described by the low slope of the *I–V* curve, which is associated with a low conductance of the nanochannel. Raising the temperature above the lower critical solubility temperature (LCST) undergoes a sharp change in the PNIPA brushes' conformational state, which suffers a transition into a collapsed state, leading to an increase of the effective cross-section of the nanochannel (Fig. 1.24a). The conformational transition into a more compact state promotes the widening of the nanochannel, which is evidenced as an increase in conductance, as derived from the slope of the *I–V* plots at 40 °C (Fig. 1.24b). The thermoresponsive brush is acting as a thermally driven macromolecular gate controlling the ionic flow through the nanochannel. Moreover, this thermoresponsive nanochannel is completely reversible, due to the reversibility of the thermally trigged conformational changes of the PNIPA brush. This fact further illustrates the versatility of polymer brushes to achieve an accurate and reversible control of the topological characteristics of nanoconfined environments with dimensions comparable to biological channels.

In a similar vein, many studies have been reported that the ionic transport behavior of the nanoporous membranes directly grafting thermo-responsive hydrogels [113] or polymer brushes [22, 114, 115] is symmetric for both directions, showing no ionic rectifying properties. It was thought that the native charged groups on the inner surfaces of the nanochannels are replaced by the uncharged polymer chains before chemical modification. Wang and Jiang et al. further explored the ionic current rectification in a conically shaped single metal-polymer nanochannel modified with PNIPA [65]. This channel can rectify ionic current in response to the temperature. At temperatures lower than 34 °C, the nanochannel rectifies the ionic current. The degree of the rectification can be enhanced with the rising temperature. While at temperatures higher than 38 °C, the nanochannel complex no longer rectifies the ionic current, showing linear and ohmic ionic transport behavior. After the self-assembly of the thiol-terminated PNIPA brushes into the nanochannel, the nanochannel exhibits closed-open switching in the ionic permeability in response to the temperature (Fig. 1.25a). The ionic current measured at the lower temperature from 21 to 29 °C was about three times lower than that measured at above 38 °C. The high and low conducting states were interconnected by a sharp transition area at the temperature from ca. 34 to 38 °C. More importantly, through this chemical modification strategy, the surface charge can be retained by the anions adsorption onto the gold channel wall, which provides possibilities to achieve the ionic rectifying nanochannels. As shown in Fig. 1.25b, asymmetry in the *I*–*V* curves is observed at 25 °C, showing ionic rectifying behavior. But the *I*–*V* curves became linear when the temperature was elevated to about 40 °C, indicating that the nanochannel no longer rectifies the ionic current. The current rectifying properties have been also observed in the control experiments on a gold-coated nanochannel without PNIPA modification (Fig. 1.25c), whereas the current rectification ratio in control experiments did not show significant change in response to the temperature. Although the ionic current rectification ratio of this composite channel is not high, it can be further enhanced by potentiostatically controlling the surface potential of the gold membrane [65].

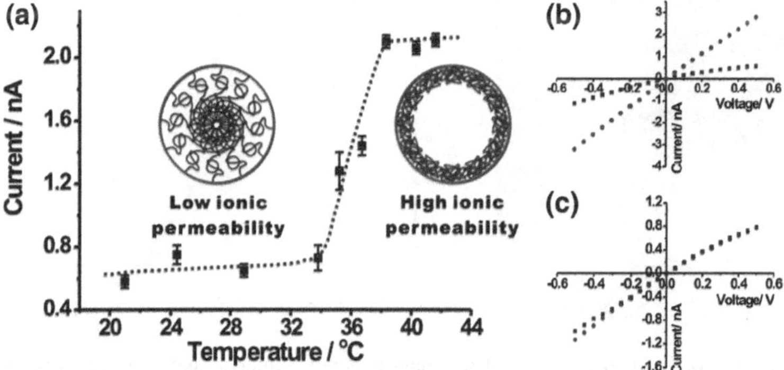

Fig. 1.25 **a** Temperature dependence of the ionic current through the PNIPA-modified single metal-polymer nanochannel. The inset cartoons illustrate the gate-like temperature responsive change in effective channel size. The applied voltage was +0.5 V. *I*–*V* curves measured on single conical gold-coated nanochannels with **b** and without **c** PNIPA modification at 25 °C (■) and 40 °C (●). Reproduced from Ref. [65] by permission of John Wiley & Sons Ltd

1.3.2.3 Light Responsive Single Nanopore/Nanochannel

Among the known addressable external stimulus, light responsive molecular switches have opened a particularly challenging and appealing research area for applications in the development of gating pores/channels for externally modulating ionic transport. It is because light response can be regulated to achieve the response of the precise region. By using structure-based design, Trauner's group have successfully developed a novel chemical gate that confers light sensitivity to an ion channel [116]. Various relevant approaches to achieve control of ions or molecules transport across nanoporous membranes have also been investigated, such as light-induced variation of nanopore diameter [117], light-induced control of the hydrophobicity of a biological nanopore [118], light control of nanopores wetting [119], and light-regulated ion transport inside nanochannels [120].

White's group reported a photon gated glass nanopore by the chemical modification of the interior surface of the nanopore with spiropyran (Fig. 1.26a), which provides a highly efficient means to impart photochemical control of molecular transport through the pore orifice [121]. In essence, the spiropyran-modified nanopore acts as a photoswitch, in which a small number of photons, absorbed by

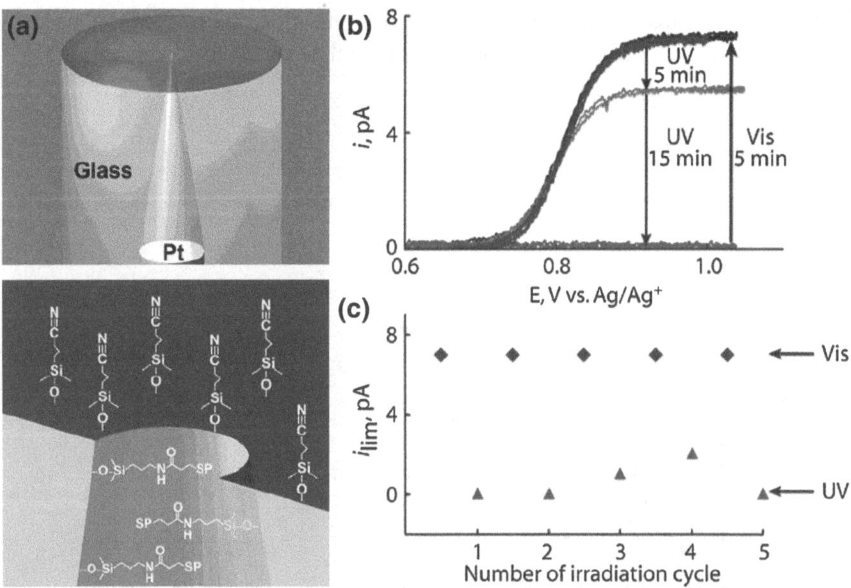

Fig. 1.26 **a** Photon gated transport at the glass nanopore. Interior and exterior surfaces of the glass nanopore modified, respectively, with spiropyran and cyanopropyl groups. **b** Voltammetric response of a spiropyran-modified glass nanopore. *Blue*, before irradiation; *magenta*, after irradiation with UV light for 5 min; *red*, after additional irradiation with UV light for 15 min; *black*, after irradiation with visible light for 5 min. **c** The magnitude of the voltammetric limiting current during five on–off irradiation cycles with UV and visible light. Reprinted with the permission from Ref. [121]. Copyright 2006 American Chemical Society

spiropyran moieties in the pore orifice, induce reversible gating of the steady-state flow of charged species through the pore orifice. The influence of UV and visible irradiation on the steady-state voltammetric response of the spiropyran modified glass nanopore was investigated, and a freshly prepared spiropyran modified nanopore with a 15 nm radius orifice, in the dark, displayed a steady-state sigmoidal-shaped voltammogram (Fig. 1.26b). Irradiation of this nanopore with 366 nm light for 5–15 min resulted in reduction of the faradaic current to background levels, and the voltammetric current can be restored by exposure to visible light for 5 min. The reversibility of transport gating at the glass nanopore was also investigated (Fig. 1.26c), and typical relatively good reproducibility was observed for 5–10 cycles, followed by a gradual decrease in the blocking efficiency upon irradiation with UV light. It was assumed that the loss of photoresponse was due to the photodegradation of spiropyran moieties.

Most recently, Zhai et al. also developed photo-induced current amplification in L-histidine modified nanochannels [122]. Figure 1.27a shows the tetravalent charged molecule can be generated under UV light. As shown in Fig. 1.27b, at a positive bias, the highly negatively charged molecule in the electrolyte solution is driven into the nanochannel by electrophoresis, resulting in a transient steric block to the smaller ions; while at a negative bias, the highly negatively charged molecule is driven out of the nanochannel by electrophoresis, which facilitates the conductance of smaller ions (Fig. 1.27c). Figure 1.27d shows UV light induced the current amplification in both directions of the bias voltage applied on the nanochannel. At negative bias, the current gradually amplifies on UV light irradiation and reverses back in darkness (Fig. 1.27e). At positive bias, the current increased (Fig. 1.27f). As a result, by UV light irradiation, they could adjust the charge of

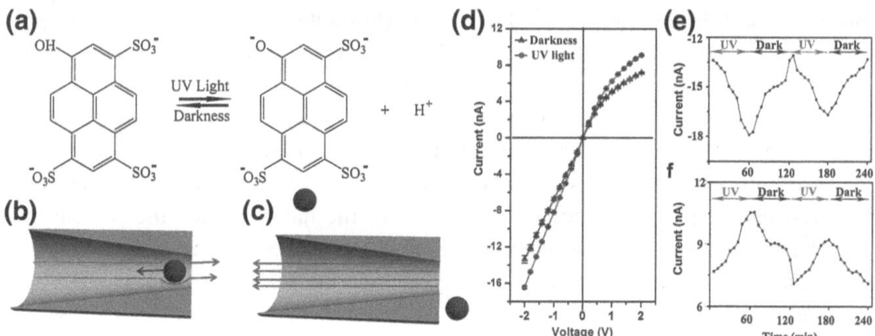

Fig. 1.27 Mechanism to explain the ionic rectification in a light-induced nanofluidic diode. **a** Photochemical reactions of the highly charged molecules under UV light. **b** At a positive bias, the highly negatively charged molecule (*blue particle*) is driven into the nanochannel, which prohibits the conductance of ionic current (*red line*). **c** At a negative bias, the highly negatively charged molecule (*blue particle*) is driven out of the nanochannel, which facilitates the conductance of ionic current (*red line*). **d** *I–V* curves on turning UV light on and off. **e** The current–time (*I–t*) curve on turning UV light on and off at −2 V. **f** *I–t* curve on turning UV light on and off at 2 V. Reproduced from Ref. [122] by permission of John Wiley & Sons Ltd

bistable photoacid molecules reversibly, therefore the photo-induced current amplification could be achieved, which offers a way to control ionic transport inside the nanochannels by turning the UV light on and off, alternately.

1.3.2.4 Electric Potential Responsive Single Nanopore/Nanochannel

A transmembrane potential is a particularly useful stimulus as it is non-invasive, tunable, and can act over a short time scale [8]. Electric potential responsive materials in this chapter exclude the original asymmetric polymer nanochannels that shows voltage-dependent ionic current fluctuations with opening and closing kinetics similar to voltage-gated ion channels [87] and biological nanopores [123]. It is clear that ionic current rectification in ion channels is more complicated and involves physical movement of an ionically charged portion of the channel in response to a change in the transmembrane potential [64].

With the inspiration of examples from nature, the scientific community started to build up biomimetic electric potential responsive channels. As in the case of Martin and Siwy et al., who created artificial ion channels that were designed to rectify the ionic current flowing through them via this "electromechanical" mechanism [64]. These artificial channels are based on conical gold single nanochannels with the critical electromechanical response provided by single-stranded DNA molecules attached to the nanochannel walls. Before DNA modification, the gold nanochannel do not rectify, even though there is adsorbed Cl^- on the nanochannel walls (Fig. 1.28a). The DNA-containing nanochannels rectify the ionic current, and it shows that an on-state at negative transmembrane potentials (anode facing the mouth of nanochannel) and an off-state at positive potentials. It was supposed that the ionic rectification in these nanochannel entailed electrophoretic insertion of the DNA chains into (off-state) and out of (on-state) the nanochannel mouth. The off-state is obtained because when inserted into the mouth, the chains partially occlude the pathway for ion transport, yielding a higher ionic resistance for the nanochannel. This hypothesis can be supported by the observation that *I–V* curves as the magnitudes of the on-state currents in Fig. 1.28a decrease with increasing DNA chain length. This is because even in the on-state, the DNA chains partially occlude the mouth of the nanochannel and increase the nanochannel resistance. Karhanek et al. also developed a poly-L-lysine coated glass nanopore, which could dramatically change the ionic current response to an identical applied voltage [24].

Recently, another single conical nanochannels were developed by Siwy's group with tunable ionic rectification properties achieved by a metal gate situated at the narrow entrance of the channels [20]. The gate voltage switches the characteristic of the channels from a rectifier to a symmetric *I–V* curve. This ionic current modulation is obtained with gate voltages lower than 1 V. Figure 1.28b shows a series of *I–V* curves of a single conical nanochannel with metal and silica layers deposited on the side with the small channel opening. Without the gate, the single conical nanochannels in pH 8 are cation selective and rectify the current such that

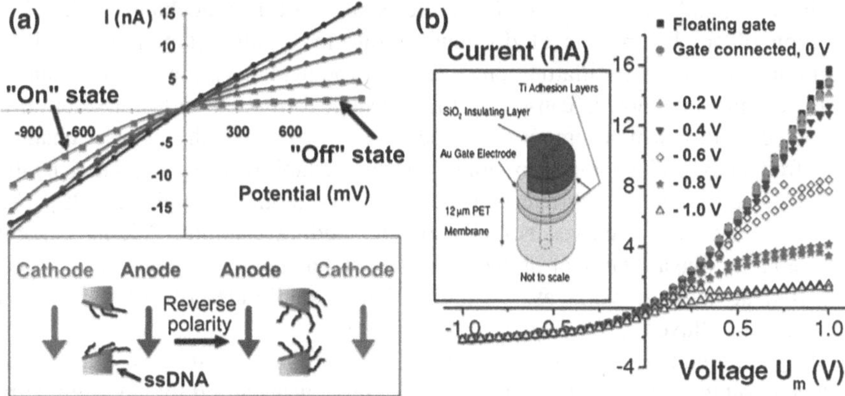

Fig. 1.28 Electric potential responsive nanochannels. **a** *I–V* curves for the DNA-modified single metal-polymer nanochannel containing no DNA (*black*) and attached 12-mer (*blue*), 15-mer (*red*), 30-mer (*green*), and 45-mer (*orange*) DNAs. Schematics showing electrode polarity and DNA chain positions for on and off states. Reprinted with the permission from Ref. [64]. Copyright 2004 American Chemical Society. **b** *I–V* curves of a single conical nanochannel with a gate placed at the narrow opening of the channels. The inset shows scheme of the nanochannel with metal and silicon dioxide layers. U_m is the voltage applied across the membrane. In the legend, the gate voltages U_{gate} are given. The measurements were performed in 0.1 M KCl, pH 8. Reprinted from Ref. [20], with kind permission from Springer Science+Business Media

the preferential direction of cation flow is from the narrow opening to the wide entrance of the channel. Thus, there are larger currents for positive voltages than currents recorded for negative voltages. Applying negative gate voltage resulted in diminishing the ionic rectification properties of the conical nanochannel, and almost symmetric *I–V* curves were obtained. The mechanism of this current-tuning involves changes in the ionic distributions at the channel entrance so that concentration polarization occurs.

1.3.2.5 Ions Responsive Single Nanopore/Nanochannel

Metallic ions are crucial in modulating the activity of muscles and nerves, cells of which have specialized ion channels for transporting these ions, such as sodium, calcium and potassium. Normal body function extremely depends on the regulation of metallic ions concentrations inside the ion channels within a certain range. For life science, ions responsive single nanopore/nanochannel materials promote a potential platform to study and simulate these processes happening in living organisms with a convenient artificial system.

Siwy et al. discovered the calcium-induced voltage gating in an original asymmetric PET single nanochannel [124], and found the channel produced voltage-dependent ionic current fluctuations in the presence of sub-millimolar concentrations of calcium ions. Millimolar concentrations of calcium reverse the rectification, and a negative incremental resistance of this nanochannel [125]. The addition of small amounts of divalent cations to a buffered monovalent ionic solution results in an oscillating ionic current through an asymmetric nanochannel. This new phenomenon is caused by the transient formation and redissolution of nanoprecipitates, which temporarily block the ionic current through the channel. The frequency and character of ionic current instabilities are regulated by the potential across the membrane and the chemistry of the precipitate [126]. Figure 1.29a shows *I–V* curves before and after adding calcium. Adding calcium induced a nonlinear behavior in the *I–V* curve, and negative incremental resistance occurred at negative voltages. Therefore, larger amplitudes of voltage induced smaller ionic fluxes.

As mentioned above, the divalent cations responsive property of calcium-induced voltage gating single conical nanochannel was caused by the asymmetric shape and the surface charge in confined space. Jiang et al. further developed a biomimetic potassium responsive single nanochannel which had an ion concentration effect that provided a nonlinear response to potassium ion at the concentration ranging from 0 to 1500 μM [25]. G-quadruplex (G4) DNA was immobilized onto the inner surface of a single nanochannel, which underwent a potassium-responsive conformational change and then induces the change in the effective size at the narrowest point of the channel. The responsive ability of this system can be regulated by the stability of G-quadruplex structure through adjusting potassium concentration. A positive correlation between Li^+ concentration and the current before and after G4 DNA modification means that the current increased with Li^+ concentration from 0 to 1500 μM (Fig 1.29b). Whereas

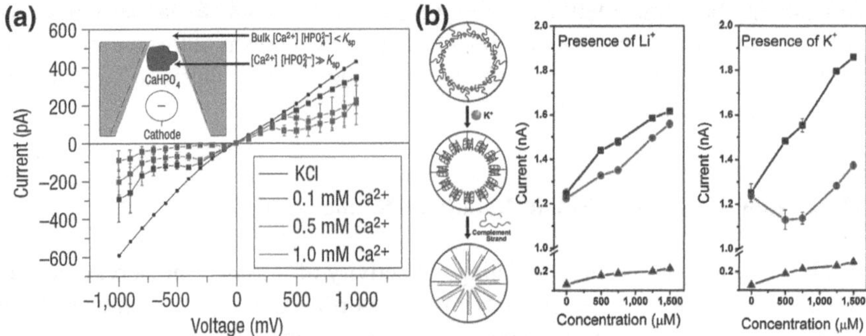

Fig. 1.29 Ion responsive nanochannels. **a** Calcium induced voltage gating in a single conical nanochannel. *I–V* curves were recorded in 0.1 M KCl with 2 mM PBS buffer alone (*black trace*) and with varying calcium concentrations. The inset shows the activities of Ca^{2+} and HPO_4^{2-} ions inside the nanochannel with negative surface charges then rise above the solubility product K_{sp} of $CaHPO_4$, allowing nanoprecipitation to occur. Reprinted by permission from Macmillan Publishers Ltd: Ref. [126], copyright 2008. **b** G-quadruplex DNA was immobilized onto the inner surface of a single nanochannel for a biomimetic potassium responsive nanochannel. Current-concentration (Li^+ and K^+) properties of the single nanochannel before and after DNA molecules attached onto the channel wall in Tris-HCl (5 mM, pH 7.2, at 23 °C). Reprinted with the permission from Ref. [25]. Copyright 2009 American Chemical Society

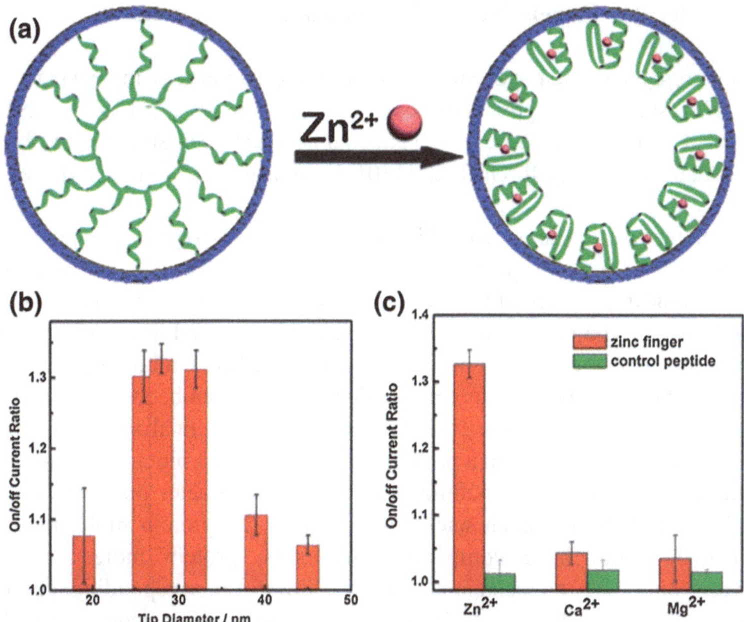

Fig. 1.30 **a** A biomimetic zinc activated nanochannel. **b** On/off current ratios measured with different tip diameters. **c** On/off current ratios measured with different ions. Reproduced from Ref. [127] by permission of The Royal Society of Chemistry

the hybridization of the complementary DNA strands with G4 DNA resulted in a sharp decrease in the currents, it still kept an increasing trend. Before G4 DNA modification and after the hybridization, the current change of the nanochannel via the concentration at different states upon the addition of K^+ indicated an increasing trend similar to Li^+. There was a remarkable difference after DNA modification for K^+. This nonlinear response phenomenon could be attributed to the formation of G4 structures that induced the relatively dense packing of DNA molecules on the inner wall of the nanochannel, resulting in an efficient decrease of the effective size and thus the current change. These results coincided with previous studies on the four-stranded i-motif structure [56]. However, the hybridization of the complementary DNA strands with G4 DNA formed the rigid duplex structure of DNA and thus created a closely packed arrangement of double-stranded DNA structure that was more stable than the G4 structure (Fig 1.29b, left cartoon). Therefore, G4 DNA conformation could not change with the increase of K^+ concentration.

Recently, a biomimetic zinc activated nanochannel (Fig. 1.30) was obtained by immobilizing zinc finger peptides into a single asymmetric nanochannel [127]. Similar to the zinc activated ion channel [128], this nanochannel is activated by zinc, which contributes to the conformational changes of the zinc fingers. Compared with simple responsive DNA molecules, it is more complicated and moves one step farther for the development of "smart" nanochannel materials.

1.3.2.6 Molecular Responsive Single Nanopore/Nanochannel

Molecular responsive nanopore/nanochannel materials attract increasing interest, one of their significant applications is molecular sensors [7, 10, 13, 129]. Biosensing with various nanopores and nanochannels has been summarized by Jiang [129], Arben Merkoçi [10], Griffiths [130], Gyurcsanyi [39], and Martin et al. [7, 12, 131].

The α-hemolysin protein channel is most used as a biological nanopore, and the sensor consists of a single pore embedded within a lipid bilayer membrane. This simple responsive property of this pore is based on size effect, that an ionic current is passed through the pore, and different molecules are used as transient blocks in order to influence the ionic transport properties [132]. Unlike the fragile lipid bilayer membrane in which most protein channels are embedded, polymer films are mechanically, chemically robust and easily functionalized. For instance, Martin and Siwy et al. reported a protein analyte binds to a biochemical molecular-recognition agent (MRA) immobilized at the small diameter opening of a single conically shaped gold nanochannel [76]. Because the protein molecule and the nanochannel mouth have comparable diameters, protein molecules binding effectively plug the nanochannel, which shows a corresponding responsive blockage of the ionic current. *I–V* curves for the biotinylated nanochannel before and after exposure to a solution containing 100 nM lysozyme are identical (Fig. 1.31), indicating that the channel does not respond to a protein that does not bind to the biotin MRA. In contrast, the ionic current is completely shut off after exposure to a solution which contains 180 pM streptavidin as analyte protein (Fig. 1.31). Total ionic current blockage occurs because the diameter of the streptavidin molecule, about 5 nm, is comparable to the mouth diameter for the biotinylated nanochannel.

Recently, Li et al. reported a simple enantioselective sensing device based on a single artificial *β*-cyclodextrin-modified nanochannel system. It shows highly selective recognition of histidine enantiomers through monitoring of ionic current signatures [60]. Figure 1.32a shows the fabrication and operating principle of the chiral-responsive system. Figure 1.32b shows the current change ratios for the

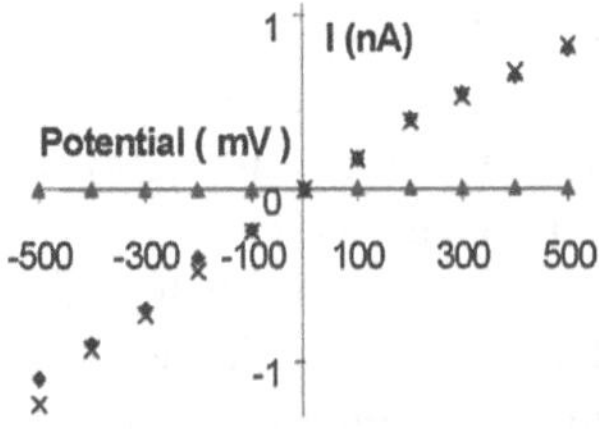

Fig. 1.31 Protein responsive conical gold nanochannels. *I–V* curves for the streptavidin responsive nanochannel in the presence of no protein (×), 100 nM lysozyme (◆), and 180 pM streptavidin (▲). Reprinted with the permission from Ref. [76]. Copyright 2005 American Chemical Society

modified nanochannel in the presence of L- or D-His, L- or D-Phe, and L- or D-Tyr. It is obvious that the β-CD-modified nanochannel displayed Hisspecific chiral discrimination. The functionalized nanochannel exhibited good chiral recognition capability toward L-His, which was manifested via the changes in the ionic current flowing through the nanochannel. Upon exposure of the β-CD-modified nanochannel to a solution of L-His, selective binding of L-His to the channel wall occurred inside the confined geometry. This effect induced a decrease in the transmembrane ionic current. In contrast, no significant changes in the ionic current were found when the modified channel was exposed to solutions of D-His or other aromatic amino acids.

1.3.3 Multiple Responsive Single Nanopore/Nanochannel

Simple responsive nanopore/nanochannel materials have been well developed, however, how to endow these pores/channels with more intelligence is still a challenging task. Several gating membranes using multiple stimuli-responsive materials have been reported, and their gates act in response to simple or multiple ambient stimuli, such as signal-responsive gating of porous membranes [133], ion and temperature dual responsive ionic gating membranes [95] and light and temperature dual responsive nanoporous membranes [117].

As already noted, the chemical properties and shape of the nanopores/nanochannels are two key factors to control ionic transport properties inside the pores/channels. According to these two factors, we suggest two strategies for designing multiple responsive nanopore/nanochannel materials. The first strategy focuses on the design and synthesis of functional molecules with multiple responsive properties on the inner surfaces of the nanopore/nanochannel materials. The second one is to prepare various symmetric and asymmetric shaped nanopores/nanochannels for different chemical modification approaches to functionalize diverse specific local areas precisely with different functional molecules. Some examples are mentioned in the following sections.

1.3.3.1 pH and Temperature Dual Responsive Single Nanopore/Nanochannel

According to above first strategy, Ulbricht et al. developed symmetric multiple nanochannels responding to pH and temperature stimuli via modifying dual responsive copolymer brushes [134, 135]. Recently, Wang and Jiang et al. presented an integrating ionic gate and rectifier within an asymmetric single nanochannel via modifying pH and temperature dual responsive copolymer brushes (Fig. 1.33a) [136]. The thermal gating ratios of this nanochannel remain approximately the same in the pH range from 3.6 to 9.4, but the ionic current rectification

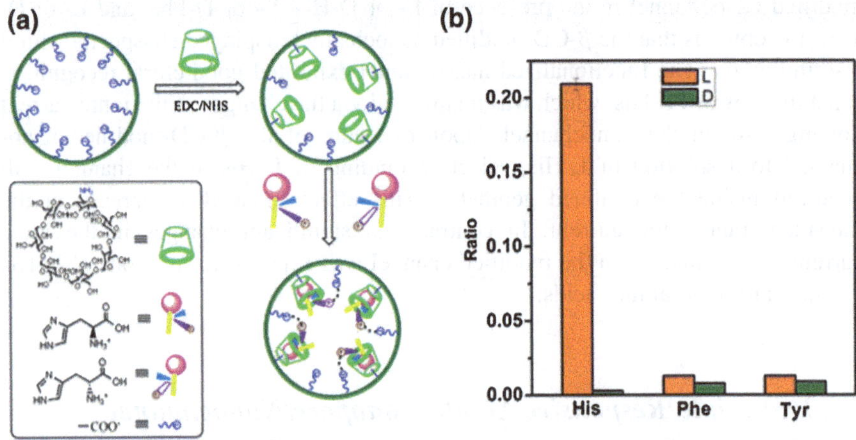

Fig. 1.32 **a** The fabrication of β-CD-modified single nanochannel and proposed mechanism for chiral recognition. **b** Current change ratio of β-CD-modified single nanochannel in 50 mM PBS with adding 1 mM L/D-His, L/D-Phe and L/D-Tyr, respectively. The sensor exhibits a remarkable selectivity and specificity toward L-His. Reprinted with the permission from Ref. [60]. Copyright 2011 American Chemical Society

ratios increase with pH, which show a trend to reach saturation at high pH (Fig. 1.33b). Moreover, the temperature-controlled ionic gate works properly in the pH range from 3.6 to 9.4 and the pH-controlled ionic rectifier works equally well at both low (25 °C) and high (40 °C) temperatures.

Based on the second strategy, a biomimetic asymmetric dual responsive single nanochannel was created by using asymmetric chemical modification approach inside the single nanochannels. (Fig. 1.33c) [136]. It displays the advanced feature of providing control over pH and temperature cooperation tunable asymmetric ionic transport property. In this dual responsive single nanochannel, there is a negative correlation between the ionic current rectification ratio and the temperature with various pH (Fig. 1.33d), while the ratio of the nanochannel before modification stayed nearly 1 at different temperatures and pH values.

1.3.3.2 pH and Molecular Dual Responsive Single Nanopore/Nanochannel

A pH and molecular dual responsive single nanochannel was created for biosensors [85]. Firstly, the original single conically shaped polymer nanochannel was modified by a recognition agent only at the narrow opening, using the method of surface patterning mentioned before [38, 84]. As the next step, the tip of this single conical nanochannel was modified with a monoclonal antibody (mAb) for the capsular the poly-γ-glutamic acid (γDPGA). The resulting nanochannel showed a

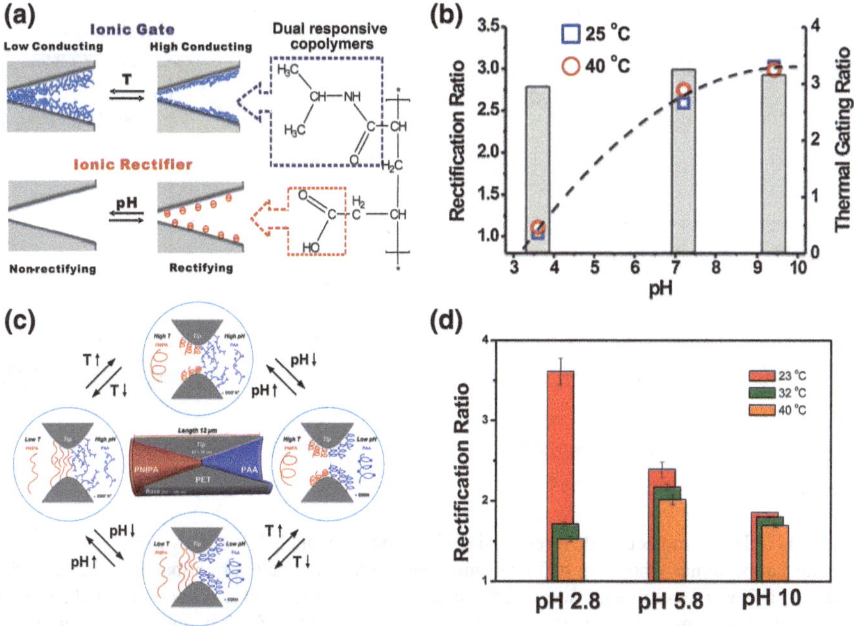

Fig. 1.33 pH/temperature dual responsive nanochannels. **a** Schematic illustration of the chemical modification of the single cone-shaped PI nanochannels with poly(*N*-isopropyl acrylamide-co-acrylic acid) copolymer brushes. **b** Ionic current rectification ratio and the thermal gating ratio measured in electrolytes of different pH. Reproduced from Ref. [136] by permission of John Wiley & Sons Ltd. **c** A biomimetic asymmetric dual responsive single nanochannel system. **d** Ionic current rectification of the single nanochannel after asymmetric chemical modification at 2 V. Reprinted with the permission from Ref. [83]. Copyright 2010 American Chemical Society

very strong dependence of its *I–V* behavior on pH (Fig. 1.34a). At pH 8, the rectification direction was the same as for an unmodified channel. Placing this channel in an acidic solution resulted in a reversed rectification and a significant increase of the rectification degree, providing evidence for the formation of a bipolar diode junction. After the channel was incubated with the bacterial γDPGA, it showed that this channel rectified ionic current only in one direction for all examined pH conditions (Fig. 1.34b). However, pH values change the magnitude of ionic currents especially for negative voltages. Obviously, the influence of the surface charge on the tip of the conical nanochannel is the key factor for regulating ionic transport properties. In a similar vein, Ali et al. also reported a functionalized single asymmetric polymer nanochannel for biosensing [61].

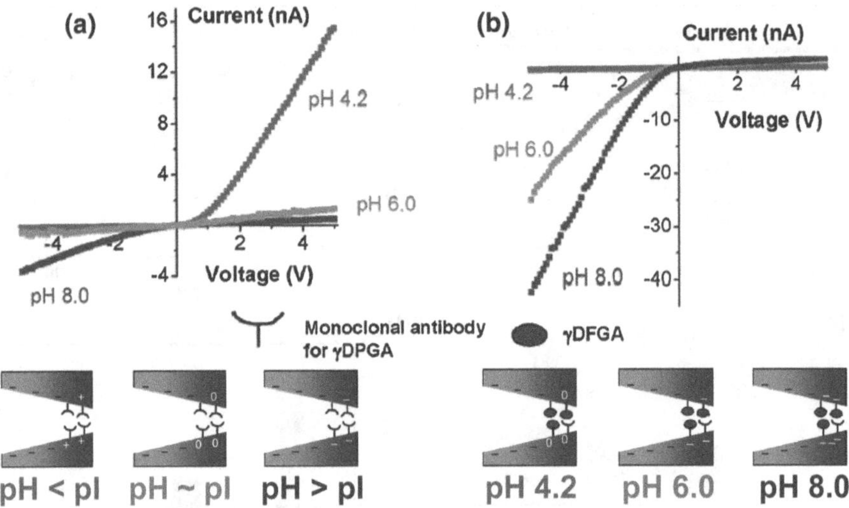

Fig. 1.34 pH and molecular dual responsive single nanochannel. *I–V* curves were recorded for a conical nanochannal whose tip had been modified with a monoclonal antibody to the bacterial γDPGA. **a** The measurements were performed in 10 mM KCl and at various pH values. **b** A nanochannel with mAb was incubated with a solution of γDPGA for 3 h, and the measurements were performed in 10 mM KCl and at various pH values. Reprinted with the permission from Ref [85]. Copyright 2009 American Chemical Society

1.4 Potential Applications of Smart Nanopore/ Nanochannel Materials

The term "application" in the context of smart nanopore/nanochannel materials does not refer to commercial products available in the market, which can only be envisioned in the long term. At present, developing smart nanopore/nanochannel materials is an interesting scientific challenge with promising applications in biosensors, nanofluidic devices, molecular filtration, and many other areas.

1.4.1 Biosensors

Biosensors of nanopore/nanochannel materials are expected to have major impact on bioanalysis and fundamental understanding of nanoscale chemical interactions down to the single molecule level. Because the major advantage of nanopores/ nanochannels offers the prospect of examining the sample in the extremely small volume defined by their interior [130]. The interest in developing various potential applications of nanopores/nanochannels has mainly been triggered by the possible application for biosensing.

Especially, a single nanochannel presents an optimal system for studying transport properties of different ions or molecules in confined space because one can observe directly the behavior of a single channel without having to average the effects of multiple channels. The first experiments to use this idea involved the biological nanopore α-hemolysin inserted in a lipid bilayer for the detection of single DNA polynucleotides [28]. Recently, there has been significant interest in the development of artificial nanochannels as sensing elements for chemical and biochemical sensors [7, 39, 85, 137], due to their mechanical and chemical stability. They enable single-molecule detection and analytical capabilities that are achieved by the measurement of ionic current blockades caused by different conformations of the molecules through a nanoscale pore or channel. It is expected that the chemistry, size, and conformational states of biomolecules passing through the channel will be reflected in the duration and magnitude of the current.

This approach has been used to investigate a wide range of biomolecular translocation events through artificial nanochannels [6, 138, 139], which has led to an increasing understanding of basic physical translocation processes. Moreover, artificial nanochannels provide two potential advantages for biosensing. First, there are numerous interactions between the ions or biomolecules in solution and the targeting molecules on the channel walls when the solution travels along the length of the nanochannel; thus, it is possible to improve signal detection [140] and achieve different quantitative analyses [141] simply by lengthening the channel. The second advantage is that the size, length, and shape of the confined space provided by nanochannels can be controlled precisely by different approaches, such as metered penetration [139] and ion-track-etching technology [85], thereby providing closer resemblance to physiological conditions. Furthermore, grafting of biomolecules on the inner walls of the nanochannel can mimic in vivo conditions (Fig. 1.35, bottom right). For example, the G-rich telomere overhang is attached to the chromosome in confined space in vivo, this condition can be more closely mimicked in an artificial nanochannel, compared to experiments performed in solution [25]. In this way, nanochannels can be used to study conformational changes of biomolecules in confined spaces, while previous nanopores cannot, because biomolecular transport in nanopores is not limited in confined spaces [138]. It is worth mentioning that conformational changes may be observed in real-time. Because the changes in the detected current are caused by the physical blockage of different biomolecular conformational states and the distribution of charge density of biomolecules on the interior surfaces of the nanochannels.

In addition, a simple strategy exists based on specific biomolecular ligands bound to recognition sites on the interior surface of nanochannels, which can then be used as recognition elements for developing a biosensor (Fig. 1.35, bottom right). By designing specific recognition sites, it may be simple to detect various biomolecules by measuring the current drop at a constant potential. This strategy is different from a resistive-pulse nanochannel sensor [132], in which the analyte is identified by the currentpulse signature. Because biomolecular analytes are comparable in size to the narrowest of nanochannels, binding of the biomolecules leads to the blocking of the nanochannel, which is detected as a permanent blockage of

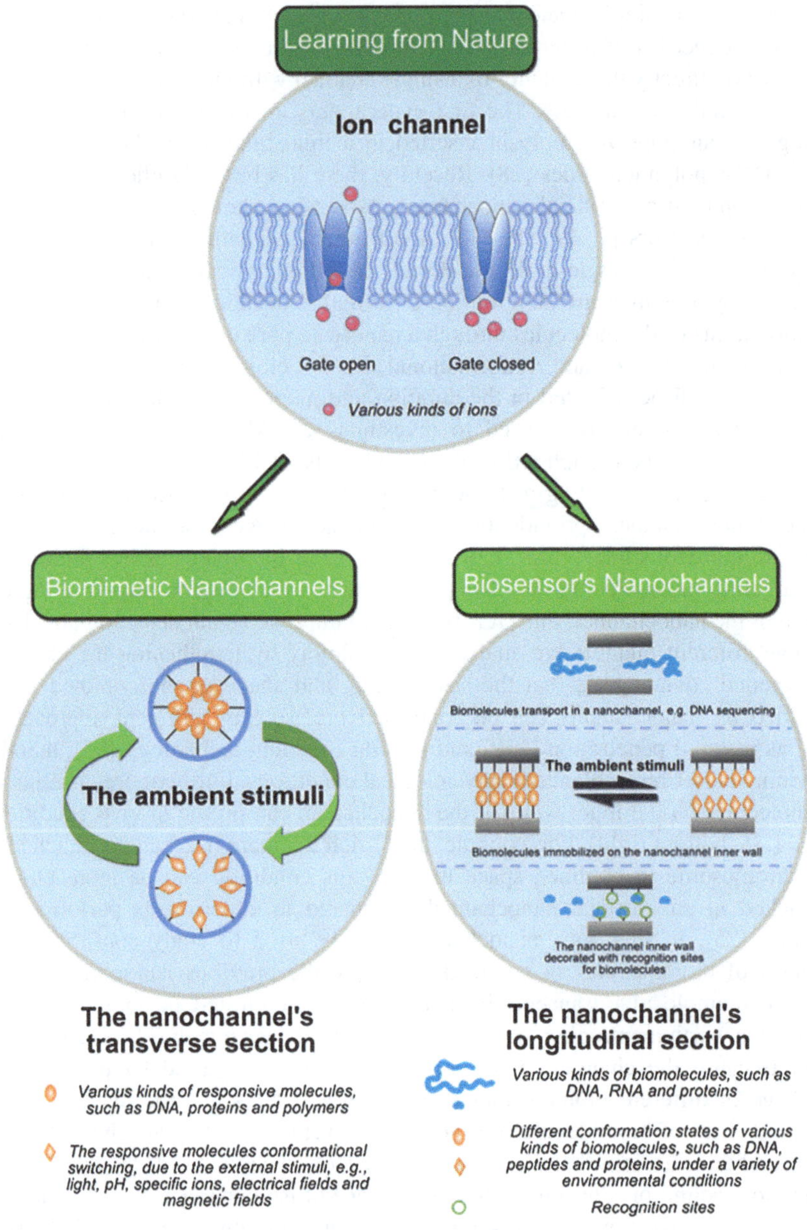

Fig. 1.35 Bio-inspired nanochannel systems: learning from nature means taking ideas from nature and developing novel functional materials based on these concepts. *Top* Inspired by natural phenomena, we can design smart artifical nanochannel systems for life science research. *Bottom left* We can simulate the process of ion transport in living organisms by using the biomimetic nanochannels. *Bottom right* Artificial nanochannel systems can be used to investigate the chemical, structural property, and conformational changes of biomolecules in confined spaces. Reprinted with the permission from Ref. [13]. Copyright 2009 American Chemical Society

the ionic current [18]. Although this specific configuration is a "one-time" sensor, if the ligands and recognition sites are only weakly bound, the sensor can be regenerated and used again [76].

For sensing with nanochannels, there are several existing challenges, which limit their stability, sensitivity, reliability, and practicability. Composite nanochannel materials, such as metal polymer compounds, provide a potential way to increase the robustness of nanochannels [20]. Another necessary consideration is the fabrication of nanochannels with high length-to-diameter ratios, which contributes to the observed translocations and higher temporal resolution [137]. Material contamination is also a significant issue for improving the cycling properties of the sensors and lowering application costs.

Although research toward bio-inspired smart nanochannels is still in its early stages, it will be enhanced greatly by the development of both chemistry and nanotechnology methods that are capable of producing more smart functional molecules and variations in the chemical and physical properties of the confined channel space. A breakthrough in biosensing for real-world applications is therefore expected from implementing artificial nanochannels, an area that is still under rapid development.

1.4.2 Nanofluidic Devices

Nanofluidic devices represent a new regime in the study of ion transport in the extremely small three dimensions [142]. There is increasing interest in measuring and investigating transport and electrochemical phenomena in nanopore/nanochannel materials. Dekker and Wang et al. reported power generation by pressure-driven transport of ions in nanofluidic channels [143, 144]. Liu et al. showed a fuel cell based on nanochannel induced proton conduction enhancement [145]. Controlling ion transport in ion channels is used throughout living systems for electrical signaling in nerves, muscles, and synapses. Inspired by the light-driven cross-membrane proton pump of biological systems, a photoelectric conversion system based on smart gating proton-driven nanochannel has been constructed [15, 91] (Fig. 1.36). Another example is inspired by the electric eel who has the inherent skill to generate considerable bioelectricity from the salt content in their body fluid with highly selective ion channels and pumps on their cell membrane. A fully abiotic single channel nanofluidic energy harvesting system that efficiently converts Gibbs free energy in the form of salinity gradient into electricity has also been created [146, 147] (Fig. 1.37).

Most recently, Jiang et al. developed a smart gating hydroxide ion-driven nanofluidic channel device to achieve a photoelectric conversion system which can work in alkaline condition based on photo-induced reversible pH changes malachite green carbinol base (MGCB) [148]. The photoelectric conversion system is composed of three main components (Fig. 1.38a–c): (1) a photoelectrochemical cell that can be irradiated from one side with platinum electrodes; (2) a

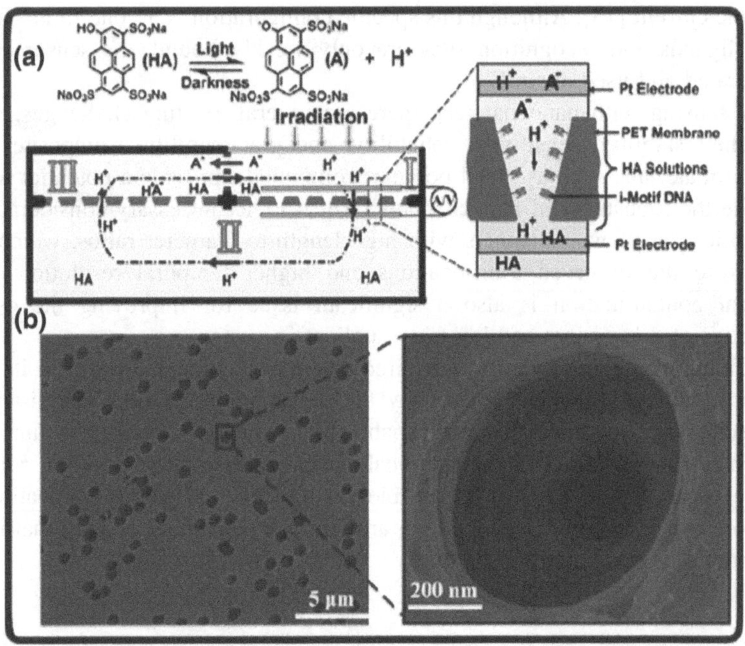

Fig. 1.36 **a** Scheme of the photoelectric conversion system, which was constructed with a photoeclectrochemical cell containing three parts and only part I could be irradiated from the outside light. **b** SEM images of the nanochannels from the base side. Reproduced from Ref. [91] by permission of John Wiley & Sons Ltd

functionalized nanochannel membrane that was prepared by grafting amino terminated groups onto the inner surface of an ion-track-etched conical nanochannel; and (3) a light-induced hydroxide ion emitter, molecular malachite green carbinol base. The response properties of the photoelectric conversion measured under irradiation and in darkness were shown in Fig. 1.38d, e.

1.4.3 Molecular Filtrations

The nanopore/nanochannel materials are also highly interesting for separation processes as molecular filtrations. Recognition of small organic molecules and large biomolecules such as proteins is of great importance in pharmaceutical as well as biological applications. Recognition inside a nanoporous membrane is particularly attractive because of the advantages associated with ligand-receptor interactions in confined spaces. Classical nanoporous membrane-based separations simply use the difference in size of the analytes relative to pore size in the membrane [149]. The density of pores/channels, their shapes, and their surface chemistry are the key factors that determine substance transport and separation

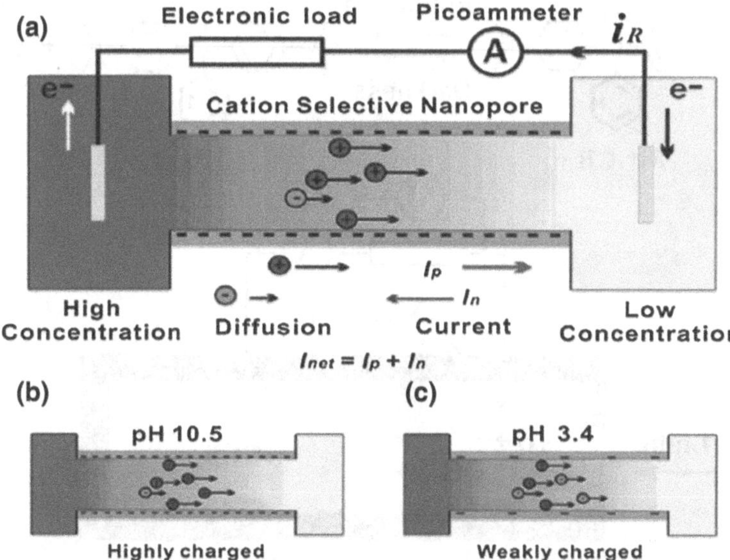

Fig. 1.37 Preferential ion diffusion across single ion-selective nanochannels driven by its concentration gradient. **a** Schematic illustration of the net diffusion current (I_{net}) generation. **b** For the highly charged nanochannel, cations are the dominant diffusive ions, resulting in high net diffusion current. **c** For the weakly charged nanochannel, the number of cations and anions are nearly identical, resulting in low net diffusion current. Reproduced from Ref. [146] by permission of John Wiley & Sons Ltd

capabilities [150]. For the purpose of selectivity beyond size, it is necessary to have methods for chemically modifying the pore/channel walls. Thayumanavan et al. reported a simple approach to functionalize the nanopores using self-assembling and non-self-assembling polymers (Fig. 1.39), and these modified nanopores separated small molecules which could be differentially transported through the nanopores based on their size and/or electrostatics [149].

1.5 Contents of the Thesis

Life through millions of years of evolution almost completed all the complex processes of intelligent control in nature. Learning from nature is the eternal theme of the development of novel smart materials and new intelligent systems. Biological nanochannels, for example, ion channels, play a very important role in the cellular basic molecular biological processes. By using nanotechnology, molecular biology, interface chemistry, statistical physics and other comprehensive research methods, a class of important basic research and a good prospect of application of smart

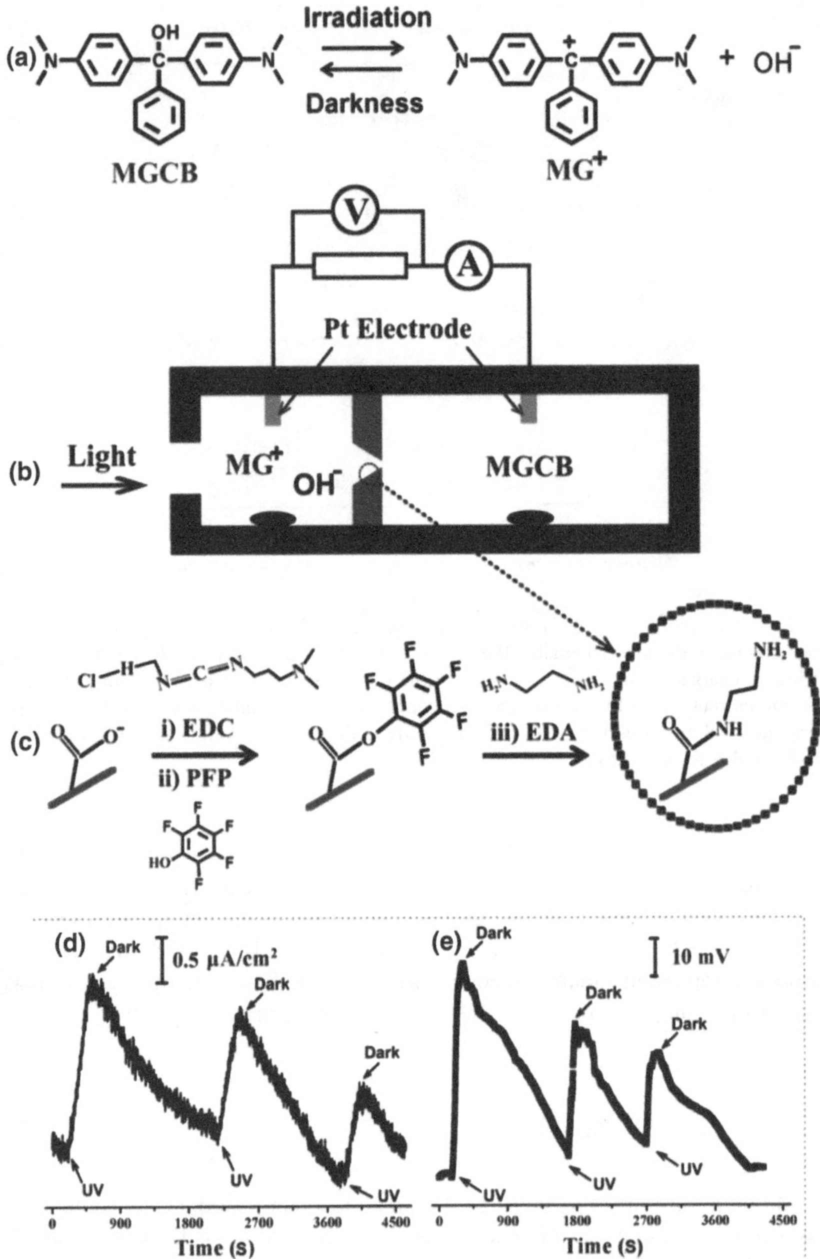

Fig. 1.38 Schematic sketch of the photoelectric conversion system. **a** Light-dependent MGCB equilibrium taking place in a photoeclectrochemical cell. **b** The experimental setup that could be irradiated from one side of the cell. **c** The chemical modification strategy for converting the carboxyl groups into amino terminated groups. **d** and **e** Photoresponse of the faradaic photocurrent and membrane potential to alternating UV irradiation and darkness. Reproduced from Ref. [148] by permission of John Wiley & Sons Ltd

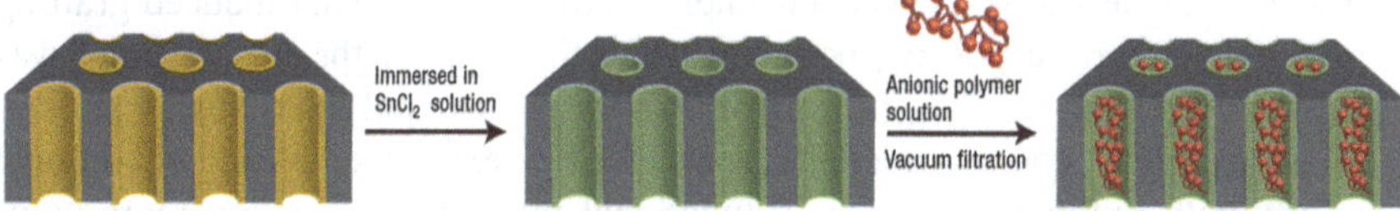

Fig. 1.39 Schematic illustration of functionalization of the nanoporous membranes with polymers. Reprinted by permission from Macmillan Publishers Ltd: Ref. [149], copyright 2008

nanopores/nanochannels will be developed. This thesis is dedicated to the study of design and development of the shapes and the functions of biomimetic single nanochannel materials. Inspired by biological nanochannels, the design and building of biomimetic smart single nanochannels can be considered from two aspects, the shape and the chemical composition of nanochannels, by using different shapes of the nanochannels, different stimuli-responsive molecules, and different asymmetric modification methods. Here we have succeeded in developing various smart single nanochannel systems to control ion transport in response to different external stimuli. It will provide the foundation for the application of functional nanochannel materials, and the primary contents of the thesis are as follows:

1. Based on the design of asymmetric conical shaped single nanochannel and K^+ responsive molecular G-quadruplex DNA sequences, we experimentally demonstrate a novel biomimetic nanochannel system which can achieve a K^+ response within a certain ion concentration range. Such a system as a basic platform could potentially spark further experimental and theoretical efforts to simulate the process of ion transport in living organisms and boost the development of bio-inspired intelligent nanochannel apparatus. Moreover, this artificial system could promote a potential to conveniently study biomolecule conformational change in confined space by the current measurement which is different from the nanopore sequencing.
2. Based on the preparation of the symmetric hour-glass shaped single nanochannel and the plasma asymmetric chemical modification approach, we experimentally demonstrate a smart nanochannel material, which displays the advanced feature of providing simultaneous control over the pH-tunable asymmetric and pH-gating ion transport properties. Before modification, the nanochannel exhibits linear *I*–*V* curves, which meant the channel did not change at different pH values and without gating property. After modification, there were significant rectifications of ionic current of the nanochannel. The open/close switching state of the channel was controlled by the pH. It closed at pH values below p*K*a, and opened at pH values above p*K*a.
3. Based on the preparation of the symmetric hour-glass shaped single nanochannel and the plasma asymmetric chemical modification approach, we experimentally demonstrate a biomimetic asymmetric responsive single nanochannel system, which displays the advanced feature of providing simultaneous control over both temperature- and pH-tunable asymmetric ionic transport

properties. One side of the nanochannel was treated by plasma-induced grafting of a temperature-responsive polymer. The other side of the nanochannel was treated by plasma-induced grafting of a pH-responsive polymer. Such a system, as an example, could potentially spark further experimental and theoretical efforts with different complicated functional molecules to exploit more complex "smart" nanochannel systems.

4. Based on the preparation of the asymmetric conical shaped single nanochannel and the ion sputtering technology, we developed the controllable continuous change of ionic current rectification of the single composite nanochannel materials with stable and chemically/structurally asymmetric properties. Compared with simple polymer materials, it is more stable and has better channel size control. Such material provides a new platform for the further self-assembled monolayer chemical modification of the inner surface of the nanochannel and the development of biosensors.

References

1. Munch E, Launey ME, Alsem DH, Saiz E, Tomsia AP, Ritchie RO (2008) Tough, bio-inspired hybrid materials. Science 322(5907):1516–1520. doi:10.1126/science.1164865
2. Xia F, Jiang L (2008) Bio-inspired, smart, multiscale interfacial materials. Adv Mater 20(15):2842–2858. doi:10.1002/adma.200800836
3. Lee H, Lee BP, Messersmith PB (2007) A reversible wet/dry adhesive inspired by mussels and geckos. Nature 448(7151):338–341. doi:10.1038/nature05968
4. Gouaux E, MacKinnon R (2005) Principles of selective ion transport in channels and pumps. Science 310(5753):1461–1465. doi:10.1126/science.1113666
5. Storm AJ, Chen JH, Ling XS, Zandbergen HW, Dekker C (2003) Fabrication of solid-state nanopores with single-nanometre precision. Nat Mater 2(8):537–540. doi:10.1038/nmat941
6. Dekker C (2007) Solid-state nanopores. Nat Nanotechnol 2(4):209–215. doi:10.1038/nnano.2007.27
7. Martin CR, Siwy ZS (2007) Learning nature's way: biosensing with synthetic nanopores. Science 317(5836):331–332. doi:10.1126/science.1146126
8. Siwy ZS, Howorka S (2010) Engineered voltage-responsive nanopores. Chem Soc Rev 39(3):1115–1132. doi:10.1039/b909105j
9. Hou X, Guo W, Jiang L (2011) Biomimetic smart nanopores and nanochannels. Chem Soc Rev 40(5):2385–2401. doi:10.1039/C0cs00053a
10. de la Escosura-Muniz A, Merkoci A (2012) Nanochannels preparation and application in biosensing. ACS Nano 6(9):7556–7583. doi:10.1021/nn301368z
11. Zhang MH, Zhai J (2012) Biomimetic smart nanoehannels for energy conversion. Prog Chem 24(4):463–470
12. Sexton LT, Horne LP, Martin CR (2007) Developing synthetic conical nanopores for biosensing applications. Mol BioSyst 3(10):667–685. doi:10.1039/b708725j
13. Hou X, Jiang L (2009) Learning from nature: building bio-inspired smart nanochannels. ACS Nano 3(11):3339–3342. doi:10.1021/Nn901402b
14. Howorka S, Siwy Z (2009) Nanopore analytics: sensing of single molecules. Chem Soc Rev 38(8):2360–2384. doi:10.1039/b813796j

15. Wen LP, Hou X, Tian Y, Nie FQ, Song YL, Zhai J, Jiang L (2010) Bioinspired smart gating of nanochannels toward photoelectric-conversion systems. Adv Mater 22(9):1021–1024. doi:10.1002/adma.200903161
16. Inglis DW, Goldys EM, Calander NP (2011) Simultaneous concentration and separation of proteins in a nanochannel. Angew Chem Int Edit 50(33):7546–7550. doi:10.1002/anie.201100236
17. Ali M, Schiedt B, Healy K, Neumann R, Ensinger A (2008) Modifying the surface charge of single track-etched conical nanopores in polyimide. Nanotechnology 19(8). http://iopscience.iop.org/0957-4484/19/8/085713/ doi:10.1088/0957-4484/19/8/085713
18. Ali M, Yameen B, Neumann R, Ensinger W, Knoll W, Azzaroni O (2008) Biosensing and supramolecular bioconjugation in single conical polymer nanochannels. Facile incorporation of biorecognition elements into nanoconfined geometries. J Am Chem Soc 130(48):16351–16357. doi:10.1021/ja8071258
19. Ali M, Yameen B, Cervera J, Ramirez P, Neumann R, Ensinger W, Knoll W, Azzaroni O (2010) Layer-by-layer assembly of polyelectrolytes into ionic current rectifying solid-state nanopores: insights from theory and experiment. J Am Chem Soc 132(24):8338–8348. doi:10.1021/ja101014y
20. Kalman EB, Sudre O, Vlassiouk I, Siwy ZS (2009) Control of ionic transport through gated single conical nanopores. Anal Bioanal Chem 394(2):413–419. doi:10.1007/s00216-008-2545-3
21. Yameen B, Ali M, Neumann R, Ensinger W, Knoll W, Azzaroni O (2009) Synthetic proton-gated ion channels via single solid-state nanochannels modified with responsive polymer brushes. Nano Lett 9(7):2788–2793. doi:10.1021/nl901403u
22. Yameen B, Ali M, Neumann R, Ensinger W, Knoll W, Azzaroni O (2009) Ionic transport through single solid-state nanopores controlled with thermally nanoactuated macromolecular gates. Small 5(11):1287–1291. doi:10.1002/smll.200801318
23. Yameen B, Ali M, Neumann R, Ensinger W, Knoll W, Azzaroni O (2010) Proton-regulated rectified ionic transport through solid-state conical nanopores modified with phosphate-bearing polymer brushes. Chem Commun 46(11):1908–1910. doi:10.1039/b920870d
24. Umehara S, Pourmand N, Webb CD, Davis RW, Yasuda K, Karhanek M (2006) Current rectification with poly-L-lysine-coated quartz nanopipettes. Nano Lett 6(11):2486–2492. doi:10.1021/nl061681k
25. Hou X, Guo W, Xia F, Nie FQ, Dong H, Tian Y, Wen LP, Wang L, Cao LX, Yang Y, Xue JM, Song YL, Wang YG, Liu DS, Jiang L (2009) A biomimetic potassium responsive nanochannel: G-Quadruplex DNA conformational switching in a synthetic nanopore. J Am Chem Soc 131(22):7800–7805. doi:10.1021/Ja901574c
26. Wendell D, Jing P, Geng J, Subramaniam V, Lee TJ, Montemagno C, Guo P (2009) Translocation of double-stranded DNA through membrane-adapted phi29 motor protein nanopores. Nat Nanotechnol 4(11):765–772. doi:10.1038/nnano.2009.259
27. Sisson AL, Shah MR, Bhosale S, Matile S (2006) Synthetic ion channels and pores (2004–2005). Chem Soc Rev 35(12):1269–1286. doi:10.1039/b512423a
28. Kasianowicz JJ, Brandin E, Branton D, Deamer DW (1996) Characterization of individual polynucleotide molecules using a membrane channel. Proc Natl Acad Sci USA 93(24):13770–13773. doi:10.1073/pnas.93.24.13770
29. Keyser UF, Koeleman BN, Van Dorp S, Krapf D, Smeets RMM, Lemay SG, Dekker NH, Dekker C (2006) Direct force measurements on DNA in a solid-state nanopore. Nat Phys 2(7):473–477. doi:10.1038/Nphys344
30. Apel P (2001) Track etching technique in membrane technology. Radiat Meas 34(1–6):559–566. doi:10.1016/s1350-4487(01)00228-1
31. Hou X, Dong H, Zhu DB, Jiang L (2010) Fabrication of stable single nanochannels with controllable ionic rectification. Small 6(3):361–365. doi:10.1002/smll.200901701
32. White RJ, Ervin EN, Yang T, Chen X, Daniel S, Cremer PS, White HS (2007) Single ion-channel recordings using glass nanopore membranes. J Am Chem Soc 129(38):11766–11775. doi:10.1021/ja073174q

33. Lathrop DK, Ervin EN, Barrall GA, Keehan MG, Kawano R, Krupka MA, White HS, Hibbs AH (2010) Monitoring the escape of DNA from a nanopore using an alternating current signal. J Am Chem Soc 132(6):1878–1885. doi:10.1021/ja906951g
34. Zhang B, Zhang YH, White HS (2004) The nanopore electrode. Anal Chem 76(21):6229–6238. doi:10.1021/ac049288r
35. Yuan JH, He FY, Sun DC, Xia XH (2004) A simple method for preparation of through-hole porous anodic alumina membrane. Chem Mater 16(10):1841–1844. doi:10.1021/cm049971u
36. Li J, Stein D, McMullan C, Branton D, Aziz MJ, Golovchenko JA (2001) Ion-beam sculpting at nanometre length scales. Nature 412(6843):166–169. doi:10.1038/35084037
37. Wu S, Park SR, Ling XS (2006) Lithography-free formation of nanopores in plastic membranes using laser heating. Nano Lett 6(11):2571–2576. doi:10.1021/nl0619498
38. Kalman EB, Vlassiouk I, Siwy ZS (2008) Nanofluidic bipolar transistors. Adv Mater 20(2):293–297. doi:10.1002/adma.200701867
39. Gyurcsanyi RE (2008) Chemically-modified nanopores for sensing. Trac-Trends Anal Chem 27(7):627–639. doi:10.1016/j.trac.2008.06.002
40. Matile S, Som A, Sorde N (2004) Recent synthetic ion channels and pores. Tetrahedron 60(31):6405–6435. doi:10.1016/j.tet.2004.05.052
41. Baker LA, Jin P, Martin CR (2005) Biomaterials and biotechnologies based on nanotube membranes. Crit Rev Solid State Mater Sci 30(4):183–205. doi:10.1080/10408430500198169
42. Huh D, Mills KL, Zhu X, Burns MA, Thouless MD, Takayama S (2007) Tuneable elastomeric nanochannels for nanofluidic manipulation. Nat Mater 6(6):424–428. doi:10.1038/nmat1907
43. Siwy Z, Apel P, Dobrev D, Neumann R, Spohr R, Trautmann C, Voss K (2003) Ion transport through asymmetric nanopores prepared by ion track etching. Nucl Instrum Methods Phys Res Sect B-Beam Interact Mater At 208:143–148. doi:10.1016/s0168-583x(03)00884-x
44. Apel PY, Blonskaya IV, Dmitriev SN, Mamonova TI, Orelovitch OL, Sartowska B, Yamauchi Y (2008) Surfactant-controlled etching of ion track nanopores and its practical applications in membrane technology. Radiat Meas 43:S552–S559. doi:10.1016/j.radmeas.2008.04.057
45. Apel PY, Blonskaya IV, Dmitriev SN, Orelovitch OL, Sartowska B (2006) Structure of polycarbonate track-etch membranes: origin of the paradoxical pore shape. J Membr Sci 282(1–2):393–400. doi:10.1016/j.memsci.2006.05.045
46. Apel PY, Korchev YE, Siwy Z, Spohr R, Yoshida M (2001) Diode-like single-ion track membrane prepared by electro-stopping. Nucl Instrum Methods Phys Res Sect B-Beam Interact Mater At 184(3):337–346. doi:10.1016/s0168-583x(01)00722-4
47. Hou X, Zhang HC, Jiang L (2012) Building bio-inspired artificial functional nanochannels: from symmetric to asymmetric modification. Angew Chem Int Ed 51(22):5296–5307. doi:10.1002/anie.201104904
48. Harrell CC, Siwy ZS, Martin CR (2006) Conical nanopore membranes: controlling the nanopore shape. Small 2(2):194–198. doi:10.1002/smll.200500196
49. Mallory GO, Hajdu JB (1990) Electroless plating: fundamentals and applications. American Electroplaters and Surface Finishers Society, Orlando
50. Nishizawa M, Menon VP, Martin CR (1995) Metal nanotubule membranes with electrochemically switchable ion-transport selectivity. Science 268(5211):700–702. doi:10.1126/science.268.5211.700
51. Menon VP, Martin CR (1995) Fabrication and evaluation of nanoelectrode ensembles. Anal Chem 67(13):1920–1928. doi:10.1021/ac00109a003
52. Meerakker JEAM (1981) On the mechanism of electroless plating. I. Oxidation of formaldehyde at different electrode surfaces. J Appl Electrochem 11(3):387–393. doi:10.1007/bf00613959

53. Li H, Lin H, Xie S, Dai W, Qiao M, Lu Y, Li H (2008) Ordered mesoporous Ni nanowires with enhanced hydrogenation activity prepared by electroless plating on functionalized SBA-15. Chem Mater 20(12):3936–3943. doi:10.1021/cm800790h
54. Krulik GA (1978) Electroless plating of plastics. J Chem Educ 55(6):361–365
55. Pernstich KP, Schenker M, Weibel F, Rossi A, Caseri WR (2010) Electroless plating of ultrathin films and mirrors of platinum nanoparticles onto polymers, metals, and ceramics. Acs Appl Mater Interfaces 2(3):639–643. doi:10.1021/am900918y
56. Xia F, Guo W, Mao YD, Hou X, Xue JM, Xia HW, Wang L, Song YL, Ji H, Qi OY, Wang YG, Jiang L (2008) Gating of single synthetic nanopores by proton-driven DNA molecular motors. J Am Chem Soc 130(26):8345–8350. doi:10.1021/Ja800266p
57. Yameen B, Ali M, Neumann R, Ensinger W, Knoll W, Azzaroni O (2009) Single conical nanopores displaying pH-tunable rectifying characteristics. Manipulating ionic transport with zwitterionic polymer brushes. J Am Chem Soc 131(6):2070–2071. doi:10.1021/ja8086104
58. Ali M, Ramirez P, Mafe S, Neumann R, Ensinger W (2009) A pH-tunable nanofluidic diode with a broad range of rectifying properties. ACS Nano 3(3):603–608. doi:10.1021/nn900039f
59. Ali M, Mafe S, Ramirez P, Neumann R, Ensinger W (2009) Logic gates using nanofluidic diodes based on conical nanopores functionalized with polyprotic acid chains. Langmuir 25(20):11993–11997. doi:10.1021/la902792f
60. Han CP, Hou X, Zhang HC, Guo W, Li HB, Jiang L (2011) Enantioselective recognition in biomimetic single artificial nanochannels. J Am Chem Soc 133(20):7644–7647. doi:10.1021/Ja2004939
61. Ali M, Schiedt B, Neumann R, Ensinger W (2010) Biosensing with Functionalized Single Asymmetric Polymer Nanochannels. Macromol Biosci 10(1):28–32. doi:10.1002/mabi.200900198
62. Ali M, Neumann R, Ensinger W (2010) Sequence-specific recognition of DNA oligomer using peptide nucleic acid (PNA)-modified synthetic ion channels: PNA/DNA hybridization in nanoconfined environment. ACS Nano 4(12):7267–7274. doi:10.1021/nn102119q
63. Wang S, Liu H, Liu D, Ma X, Fang X, Jiang L (2007) Enthalpy-driven three-state switching of a superhydrophilic/superhydrophobic surface. Angew Chem Int Ed 46(21):3915–3917. doi:10.1002/anie.200700439
64. Harrell CC, Kohli P, Siwy Z, Martin CR (2004) DNA-Nanotube artificial ion channels. J Am Chem Soc 126(48):15646–15647. doi:10.1021/ja044948v
65. Guo W, Xia HW, Xia F, Hou X, Cao LX, Wang L, Xue JM, Zhang GZ, Song YL, Zhu DB, Wang YG, Jiang L (2010) Current rectification in temperature-responsive single nanopores. ChemPhysChem 11(4):859–864. doi:10.1002/cphc.200900989
66. Hulteen JC, Jirage KB, Martin CR (1998) Introducing chemical transport selectivity into gold nanotubule membranes. J Am Chem Soc 120(26):6603–6604. doi:10.1021/ja980045o
67. Jirage KB, Hulteen JC, Martin CR (1999) Effect of thiol chemisorption on the transport properties of gold nanotubule membranes. Anal Chem 71(21):4913–4918. doi:10.1021/ac990615i
68. Chun KY, Stroeve P (2002) Protein transport in nanoporous membranes modified with self-assembled monolayers of functionalized thiols. Langmuir 18(12):4653–4658. doi:10.1021/la011250b
69. Ku J-R, Lai S-M, Ileri N, Ramirez P, Mafe S, Stroeve P (2007) pH and ionic strength effects on amino acid transport through Au-nanotubule membranes charged with self-assembled monolayers. J Phys Chem C 111(7):2965–2973. doi:10.1021/jp066944d
70. Lee SB, Martin CR (2001) pH-switchable, ion-permselective gold nanotubule membrane based on chemisorbed cysteine. Anal Chem 73(4):768–775. doi:10.1021/ac0008901
71. Lee SB, Martin CR (2001) Controlling the transport properties of gold nanotubule membranes using chemisorbed thiols. Chem Mater 13(10):3236–3244. doi:10.1021/cm0101071

72. Jagerszki G, Gyurcsanyi RE, Hofler L, Pretsch E (2007) Hybridization-modulated ion fluxes through peptide-nucleic-acid-functionalized gold nanotubes. A new approach to quantitative label-free DNA analysis. Nano Lett 7(6):1609–1612. doi:10.1021/nl0705438
73. Kim BY, Swearingen CB, Ho J-aA, Romanova EV, Bohn PW, Sweedler JV (2007) Direct immobilization of Fab' in nanocapillaries for manipulating mass-limited samples. J Am Chem Soc 129(24):7620–7626. doi:10.1021/ja070041w
74. Gyurcsanyi RE, Vigassy T, Pretsch E (2003) Biorecognition-modulated ion fluxes through functionalized gold nanotubules as a novel label-free biosensing approach. Chem Commun 20:2560–2561. doi:10.1039/b307393a
75. Jagerszki G, Takacs A, Bitter I, Gyurcsanyi RE (2011) Solid-State ion channels for potentiometric sensing. Angew Chem Int Ed 50(7):1656–1659. doi:10.1002/anie.201003849
76. Siwy Z, Trofin L, Kohli P, Baker LA, Trautmann C, Martin CR (2005) Protein biosensors based on biofunctionalized conical gold nanotubes. J Am Chem Soc 127(14):5000–5001. doi:10.1021/ja043910f
77. Alem H, Blondeau F, Glinel K, Demoustier-Champagne S, Jonas AM (2007) Layer-by-layer assembly of polyelectrolytes in nanopores. Macromolecules 40(9):3366–3372. doi:10.1021/ma0703251
78. Schmuhl R, van den Berg A, Blank DHA, ten Elshof JE (2006) Surfactant-modulated switching of molecular transport in nanometer-sized pores of membrane gates. Angew Chem Int Ed 45(20):3341–3345. doi:10.1002/anie.200504579
79. Tian Y, Hou X, Jiang L (2011) Biomimetic ionic rectifier systems: Asymmetric modification of single nanochannels by ion sputtering technology. J Electroanal Chem 656(1–2):231–236. doi:10.1016/j.jelechem.2010.11.005
80. Lau KKS, Gleason KK (2006) Initiated chemical vapor deposition (iCVD) of poly(alkyl acrylates): an experimental study. Macromolecules 39(10):3688–3694. doi:10.1021/ma0601619
81. Asatekin A, Gleason KK (2011) Polymeric nanopore membranes for hydrophobicity-based separations by conformal Initiated chemical vapor deposition. Nano Lett 11(2):677–686. doi:10.1021/nl103799d
82. Hou X, Liu Y, Dong H, Yang F, Li L, Jiang L (2010) A pH-gating ionic transport nanodevice: asymmetric chemical modification of single nanochannels. Adv Mater 22(22): 2440–2443. doi:10.1002/adma.200904268
83. Hou X, Yang F, Li L, Song Y, Jiang L, Zhu D (2010) A biomimetic asymmetric responsive single nanochannel. J Am Chem Soc 132(33):11736–11742. doi:10.1021/ja1045082
84. Vlassiouk I, Siwy ZS (2007) Nanofluidic diode. Nano Lett 7(3):552–556. doi:10.1021/nl062924b
85. Vlassiouk I, Kozel TR, Siwy ZS (2009) Biosensing with nanofluidic diodes. J Am Chem Soc 131(23):8211–8220. doi:10.1021/ja901120f
86. Wang L, Yan Y, Xie Y, Chen L, Xue J, Yan S, Wang Y (2011) A method to tune the ionic current rectification of track-etched nanopores by using surfactant. Phys Chem Chem Phys 13(2):576–581. doi:10.1039/c0cp00587h
87. Siwy Z, Apel P, Baur D, Dobrev DD, Korchev YE, Neumann R, Spohr R, Trautmann C, Voss KO (2003) Preparation of synthetic nanopores with transport properties analogous to biological channels. Surf Sci 532:1061–1066. doi:10.1016/s0039-6028(03)00448-5
88. Martin CR, Nishizawa M, Jirage K, Kang MS, Lee SB (2001) Controlling ion-transport selectivity in gold nanotubule membranes. Adv Mater 13(18):1351–1362. doi:10.1002/1521-4095(200109)13:18<1351:aid-adma1351>3.0.co;2-w
89. Alcaraz A, Ramirez P, Garcia-Gimenez E, Lopez ML, Andrio A, Aguilella VM (2006) A pH-tunable nanofluidic diode: Electrochemical rectification in a reconstituted single ion channel. J Phys Chem B 110(42):21205–21209. doi:10.1021/jp063204w
90. Tsekouras G, Johansson O, Lomoth R (2009) A surface-attached Ru complex operating as a rapid bistable molecular switch. Chem Commun 23:3425–3427. doi:10.1039/b904248b

91. Wen L, Hou X, Tian Y, Zhai J, Jiang L (2010) Bio-inspired photoelectric conversion based on smart-gating nanochannels. Adv Funct Mater 20(16):2636–2642. doi:10.1002/adfm.201000239
92. Siwy Z, Heins E, Harrell CC, Kohli P, Martin CR (2004) Conical-nanotube ion-current rectifiers: the role of surface charge. J Am Chem Soc 126(35):10850–10851. doi:10.1021/ja047675c
93. Savariar EN, Sochat MM, Klaikherd A, Thayumanavan S (2009) Functional group density and recognition in polymer nanotubes. Angew Chem Int Ed 48(1):110–114. doi:10.1002/anie.200804136
94. Li YF, Sen D (1997) Toward an efficient DNAzyme. Biochemistry 36(18):5589–5599. doi:10.1021/bi962694n
95. Yamaguchi T, Ito T, Sato T, Shinbo T, Nakao S (1999) Development of a fast response molecular recognition ion gating membrane. J Am Chem Soc 121(16):4078–4079. doi:10.1021/ja984170b
96. Siwy ZS (2006) Ion-current rectification in nanopores and nanotubes with broken symmetry. Adv Funct Mater 16(6):735–746. doi:10.1002/adfm.200500471
97. Vlassiouk I, Smirnov S, Siwy Z (2008) Ionic selectivity of single nanochannels. Nano Lett 8(7):1978–1985. doi:10.1021/Nl800949k
98. Siwy Z, Fulinski A (2002) Fabrication of a synthetic nanopore ion pump. Phys Rev Lett 89(19):198103. doi:10.1103/PhysRevLett.89.198103
99. Siwy Z, Fulinski A (2004) A nanodevice for rectification and pumping ions. Am J Phys 72(5):567–574. doi:10.1119/1.1648328
100. Woermann D (2004) Electrochemical transport properties of a cone-shaped nanopore: revisited. Phys Chem Chem Phys 6(12):3130–3132. doi:10.1039/B316166h
101. Woermann D (2003) Electrochemical transport properties of a cone-shaped nanopore: high and low electrical conductivity states depending on the sign of an applied electrical potential difference. Phys Chem Chem Phys 5(9):1853–1858. doi:10.1039/B301021j
102. Cervera J, Schiedt B, Ramirez P (2005) A Poisson/Nernst-Planck model for ionic transport through synthetic conical nanopores. Europhys Lett 71(1):35–41. doi:10.1209/epl/i2005-10054-x
103. Cervera J, Schiedt B, Neumann R, Mafe S, Ramirez P (2006) Ionic conduction, rectification, and selectivity in single conical nanopores. J Chem Phys 124(10):104706. doi:10.1063/1.2179797
104. Cervera J, Alcaraz A, Schiedt B, Neumann R, Ramirez P (2007) Asymmetric selectivity of synthetic conical nanopores probed by reversal potential measurements. J Phys Chem C 111(33):12265–12273. doi:10.1021/Jp071884c
105. Vlassiouk I, Smirnov S, Siwy Z (2008) Nanofluidic ionic diodes. Comparison of analytical and numerical solutions. Acs Nano 2(8):1589–1602. doi:10.1021/nn800306u
106. Cruz-Chu ER, Ritz T, Siwy ZS, Schulten K (2009) Molecular control of ionic conduction in polymer nanopores. Faraday Discuss 143:47–62. doi:10.1039/B906279n
107. Tagliazucchi M, Azzaroni O, Szleifer I (2010) Responsive polymers end-tethered in solid-state nanochannels: when nanoconfinement really matters. J Am Chem Soc 132(35): 12404–12411. doi:10.1021/ja104152g
108. Wei C, Bard AJ, Feldberg SW (1997) Current rectification at quartz nanopipet electrodes. Anal Chem 69(22):4627–4633. doi:10.1021/ac970551g
109. Wanunu M, Meller A (2007) Chemically modified solid-state nanopores. Nano Lett 7(6):1580–1585. doi:10.1021/nl070462b
110. Huang J, Zhang X, McNaughton PA (2006) Modulation of temperature-sensitive TRP channels. Semin Cell Dev Biol 17(6):638–645. doi:10.1016/j.semcdb.2006.11.002
111. Latorre R, Brauchi S, Orta G, Zaelzer C, Vargas G (2007) Thermo TRP channels as modular proteins with allosteric gating. Cell Calcium 42(4–5):427–438. doi:10.1016/j.ceca.2007.04.004
112. Jung Y, Bayley H, Movileanu L (2006) Temperature-responsive protein pores. J Am Chem Soc 128(47):15332–15340. doi:10.1021/ja065827t

113. Reber N, Kuchel A, Spohr R, Wolf A, Yoshida M (2001) Transport properties of thermo-responsive ion track membranes. J Membr Sci 193(1):49–58. doi:10.1016/s0376-7388(01)00460-4
114. Alem H, Duwez A-S, Lussis P, Lipnik P, Jonas AM, Demoustier-Champagne S (2008) Microstructure and thermo-responsive behavior of poly (N-isopropylacrylamide) brushes grafted in nanopores of track-etched membranes. J Membr Sci 308(1–2):75–86. doi:10.1016/j.memsci.2007.09.036
115. Lokuge I, Wang X, Bohn PW (2007) Temperature-controlled flow switching in nanocapillary array membranes mediated by poly (N-isopropylacrylamide) polymer brushes grafted by atom transfer radical polymerization. Langmuir 23(1):305–311. doi:10.1021/la060813m
116. Banghart M, Borges K, Isacoff E, Trauner D, Kramer RH (2004) Light-activated ion channels for remote control of neuronal firing. Nat Neurosci 7(12):1381–1386. doi:10.1038/nn1356
117. Liu NG, Dunphy DR, Atanassov P, Bunge SD, Chen Z, Lopez GP, Boyle TJ, Brinker CJ (2004) Photoregulation of mass transport through a photoresponsive azobenzene-modified nanoporous membrane. Nano Lett 4(4):551–554. doi:10.1021/nl0350783
118. Kocer A, Walko M, Meijberg W, Feringa BL (2005) A light-actuated nanovalve derived from a channel protein. Science 309(5735):755–758. doi:10.1126/science.1114760
119. Vlassiouk I, Park CD, Vail SA, Gust D, Smirnov S (2006) Control of nanopore wetting by a photochromic spiropyran: A light-controlled valve and electrical switch. Nano Lett 6(5):1013–1017. doi:10.1021/nl060313d
120. Zhang QQ, Liu ZY, Hou X, Fan X, Zhai J, Jiang L (2012) Light-regulated ion transport through artificial ion channels based on TiO2 nanotubular arrays. Chem Commun 48(47):5901–5903. doi:10.1039/C2cc32451b
121. Wang G, Bohaty AK, Zharov I, White HS (2006) Photon gated transport at the glass nanopore electrode. J Am Chem Soc 128(41):13553–13558. doi:10.1021/ja064274j
122. Zhang MH, Hou X, Wang JT, Tian Y, Fan X, Zhai J, Jiang L (2012) Light and pH cooperative nanofluidic diode using a spiropyran-functionalized single nanochannel. Adv Mater 24(18):2424–2428. doi:10.1002/adma.201104536
123. Maglia G, Restrepo MR, Mikhailova E, Bayley H (2008) Enhanced translocation of single DNA molecules through alpha-hemolysin nanopores by manipulation of internal charge. Proc Natl Acad Sci USA 105(50):19720–19725. doi:10.1073/pnas.0808296105
124. Siwy ZS, Powell MR, Petrov A, Kalman E, Trautmann C, Eisenberg RS (2006) Calcium-induced voltage gating in single conical nanopores. Nano Lett 6(8):1729–1734. doi:10.1021/nl061114x
125. Siwy ZS, Powell MR, Kalman E, Astumian RD, Eisenberg RS (2006) Negative incremental resistance induced by calcium in asymmetric nanopores. Nano Lett 6(3):473–477. doi:10.1021/nl0524290
126. Powell MR, Sullivan M, Vlassiouk I, Constantin D, Sudre O, Martens CC, Eisenberg RS, Siwy ZS (2008) Nanoprecipitation-assisted ion current oscillations. Nat Nanotechnol 3(1):51–57. doi:10.1038/nnano.2007.420
127. Tian Y, Hou X, Wen L, Guo W, Song Y, Sun H, Wang Y, Jiang L, Zhu D (2010) A biomimetic zinc activated ion channel. Chem Commun 46(10):1682–1684. doi:10.1039/b918006k
128. Davies PA, Wang W, Hales TG, Kirkness EF (2003) A novel class of ligand-gated ion channel is activated by Zn^{2+}. J Biol Chem 278(2):712–717. doi:10.1074/jbc.M208814200
129. Tian Y, Wen LP, Hou X, Hou GL, Jiang L (2012) Bioinspired ion-transport properties of solid-state single nanochannels and their applications in sensing. ChemPhysChem 13(10):2455–2470. doi:10.1002/cphc.201200057
130. Griffiths J (2008) The realm of the nanopore. Anal Chem 80(1):23–27. doi:10.1021/ac085995z

131. Choi Y, Baker LA, Hillebrenner H, Martin CR (2006) Biosensing with conically shaped nanopores and nanotubes. Phys Chem Chem Phys 8(43):4976–4988. doi:10.1039/b607360c
132. Sexton LT, Horne LP, Sherrill SA, Bishop GW, Baker LA, Martin CR (2007) Resistive-pulse studies of proteins and protein/antibody complexes using a conical nanotube sensor. J Am Chem Soc 129(43):13144–13152. doi:10.1021/ja0739943
133. Ito Y, Park YS (2000) Signal-responsive gating of porous membranes by polymer brushes. Polym Adv Technol 11(3):136–144. doi:10.1002/1099-1581(200003)11:3<136:aid-pat961>3.0.co;2-f
134. Geismann C, Tomicki F, Ulbricht M (2009) Block copolymer photo-grafted poly(ethylene terephthalate) capillary pore membranes distinctly switchable by two different stimuli. Sep Sci Technol 44(14):3312–3329. doi:10.1080/01496390903212755
135. Friebe A, Ulbricht M (2009) Cylindrical pores responding to two different stimuli via surface-initiated atom transfer radical polymerization for synthesis of grafted diblock copolymers. Macromolecules 42(6):1838–1848. doi:10.1021/ma802185d
136. Guo W, Xia H, Cao L, Xia F, Wang S, Zhang G, Song Y, Wang Y, Jiang L, Zhu D (2010) Integrating ionic gate and rectifier within one solid-state nanopore via modification with dual-responsive copolymer brushes. Adv Funct Mater 20(20):3561–3567. doi:10.1002/adfm.201000989
137. Mara A, Siwy Z, Trautmann C, Wan J, Kamme F (2004) An asymmetric polymer nanopore for single molecule detection. Nano Lett 4(3):497–501. doi:10.1021/nl035141o
138. Branton D, Deamer DW, Marziali A, Bayley H, Benner SA, Butler T, Di Ventra M, Garaj S, Hibbs A, Huang X, Jovanovich SB, Krstic PS, Lindsay S, Ling XS, Mastrangelo CH, Meller A, Oliver JS, Pershin YV, Ramsey JM, Riehn R, Soni GV, Tabard-Cossa V, Wanunu M, Wiggin M, Schloss JA (2008) The potential and challenges of nanopore sequencing. Nat Biotechnol 26(10):1146–1153. doi:10.1038/nbt.1495
139. Sowerby SJ, Petersen GB (2009) A proposition for single molecule DNA sequencing through a nanopore entropic trap. Int J Nanotechnol 6(3–4):398–407
140. Iqbal SM, Akin D, Bashir R (2007) Solid-state nanopore channels with DNA selectivity. Nat Nanotechnol 2(4):243–248. doi:10.1038/nnano.2007.78
141. Kohli P, Harrell CC, Cao ZH, Gasparac R, Tan WH, Martin CR (2004) DNA-functionalized nanotube membranes with single-base mismatch selectivity. Science 305(5686):984–986. doi:10.1126/science.1100024
142. Stein D, Kruithof M, Dekker C (2004) Surface-charge-governed ion transport in nanofluidic channels. Phys Rev Lett 93(3):035901. http://link.aps.org/doi/10.1103/PhysRevLett.93.035901 doi:10.1103/PhysRevLett.93.035901
143. van der Heyden FHJ, Bonthuis DJ, Stein D, Meyer C, Dekker C (2007) Power generation by pressure-driven transport of ions in nanofluidic channels. Nano Lett 7(4):1022–1025. doi:10.1021/nl070194h
144. Xie Y, Wang X, Xue J, Jin K, Chen L, Wang Y (2008) Electric energy generation in single track-etched nanopores. Appl Phys Lett 93(16):163116. doi:10.1063/1.3001590
145. Liu SR, Pu QS, Gao L, Korzeniewski C, Matzke C (2005) From nanochannel-induced proton conduction enhancement to a nanochannel-based fuel cell. Nano Lett 5(7):1389–1393. doi:10.1021/nl050712t
146. Guo W, Cao L, Xia J, Nie F-Q, Ma W, Xue J, Song Y, Zhu D, Wang Y, Jiang L (2010) Energy harvesting with single-ion-selective nanopores: A concentration-gradient-driven nanofluidic power source. Adv Funct Mater 20(8):1339–1344. doi:10.1002/adfm.200902312
147. Xu J, Lavan DA (2008) Designing artificial cells to harness the biological ion concentration gradient. Nat Nanotechnol 3(11):666–670. doi:10.1038/nnano.2008.274
148. Wen L, Tian Y, Guo Y, Ma J, Liu W, Jiang L (2013) Conversion of light to electricity by photoinduced reversible pH changes and biomimetic nanofluidic channels. Adv Funct Mater. doi:10.1002/adfm.201203259

149. Savariar EN, Krishnamoorthy K, Thayumanavan S (2008) Molecular discrimination inside polymer nanotubules. Nat Nanotechnol 3(2):112–117. doi:10.1038/nnano.2008.6
150. Vlassiouk I, Apel PY, Dmitriev SN, Healy K, Siwy ZS (2009) Versatile ultrathin nanoporous silicon nitride membranes. Proc Natl Acad Sci USA 106(50):21039–21044. doi:10.1073/pnas.0911450106

Chapter 2
Ions Responsive Asymmetric Conical Shaped Single Nanochannel

2.1 Introduction

Potassium is especially crucial in modulating the activity of muscles and nerves, cells of which have specialized ion channels for transporting potassium. Normal body function extremely depends on the regulation of potassium concentrations inside the ion channels within a certain range. For life science, undoubtedly, it is significant and challenging to study and imitate these processes happening in living organisms with a convenient artificial system. In this chapter, I introduce a novel biomimetic nanochannel system which has an ion concentration effect that provides a nonlinear response to potassium ion at the concentration ranging from 0 to 1500 μM [1]. This new phenomenon is caused by the G-quadruplex (G4) DNA conformational change with a positive correlation with ion concentration. In this work, G4 DNA was immobilized onto a synthetic nanochannel, which undergoes a potassium-responsive conformational change and then induces the change in the effective channel size. The responsive ability of this system can be regulated by the stability of G4 structure through adjusting potassium concentration. The situation of the grafting G4 DNA on a single nanochannel can closely imitate the in vivo condition because the G-rich telomere overhang is attached to the chromosome. Therefore, this artificial system could promote a potential to conveniently study biomolecule conformational change in confined space by the current measurement, which is significantly different from the nanopore sequencing. Moreover, such a system may also potentially spark further experimental and theoretical efforts to simulate the process of ion transport in living organisms and can be further generalized to other more complicated functional molecules for the exploitation of novel bio-inspired intelligent nanochannel machines.

Inspired by the biological ion channel, a synthetic film with a single nanochannel structure was prepared as described by Apel et al. [2]. Unlike the fragile lipid-bilayer membrane in which most natural ion channels are embedded, this synthetic film is mechanically and chemically robust. Recently, Martin and Siwy

X. Hou, *Bio-inspired Asymmetric Design and Building of Biomimetic Smart Single Nanochannels*, Springer Theses,
DOI: 10.1007/978-3-642-38050-1_2, © Springer-Verlag Berlin Heidelberg 2013

et al. [3, 4]. reported DNA-nanotube artificial ion channels systems which helped us to better understand the role of an electromechanical gate which responds to the applied voltage, and they also developed the function of this system with DNA single-base mismatch selectivity. J.E ten Elshof et al. reported a system using the nanometer-sized pores of membrane gates to achieve the surfactant-modulated switching of molecular transport [5]. Letant et al. reported localized functionalization of single nanopores [6]. Bashir et al. developed solid-state nanopore channels with DNA selectivity [7]. Jiang et al. reported the gating of single synthetic nanochannels by proton-driven DNA molecular motors [8]. Azzaroni and Ali et al. reported supramolecular bioconjugation in single nanochannel as a biosensor and the
pH-tunable rectifying characteristics of the single conical nanochannels [9–11]. Although ion channels in living organism have been studied by a mimic method using synthetic nanochannels during the past several decades, how to endow these synthetic nanochannels with intelligence is still a challenging task.

In this work, I further extend the function molecules-nanochannel system by using G4 DNA [12–19]. In this biomimetic nanochannel system (BNCS), there is an ion concentration effect, which is a very important phenomenon in living body while other systems do not have. This system is also different from the previous systems [3] utilized the different chain length of non-responsive DNA oligomers, which chemisorbed on the channel walls and surface of the membrane. According to the pH influences on the nanochannel, the previous work about pH responsive i-motif DNA-nanochannel system focused on the permeability ratios of the single nanochannel between two different pH values [8]. The present work is different, because this novel biomimetic nanochannel system was responsive to potassium ion (K^+) within a certain concentration range and simulated these processes in pH neutral environment as in a natural organism [20].

G-quadruplexes are highly ordered DNA structures derived from G-rich sequences formed by tetrads of hydrogen-bonded guanine bases. Among the quadruplex-forming sequences, the human telomeric sequence d[AGGG $(TTAGGG)_3$] has attracted tremendous interest due to its importance at telomere maintenance and cell aging or death [14]. Recently, there are many researches on the conformational change of this biomolecule [16, 18, 21–23] and its applications in artificial ion channels [24–29] and biosensors [30–33]. Here we selected the G-rich human telomere strand [12, 13, 22, 34] due to the possibility of forming the intramolecular four-stranded quadruplex topologies, and this process is dependent on the alkali metal ion concentration. Considering the efficiency at stabilizing G4 DNA, K^+ is much more effective than other alkali metal ions in promoting the formation of the G4 structures, whereas lithium ion (Li^+) is ineffective in all cases [16, 35]. Therefore, these two ions mentioned above were selected as control to investigate the ion responsive properties of the BNCS.

We prepared a single nanochannel membrane with the well-developed ion-track-etching technique [36–39]. The nanochannel (Fig. 1.6) was embedded in a track-etched polyethylene terephthalate membrane (PET, Hostaphan RN12 Hoechst, 12 μm thick, with single ion track in the center). The track-etching technique

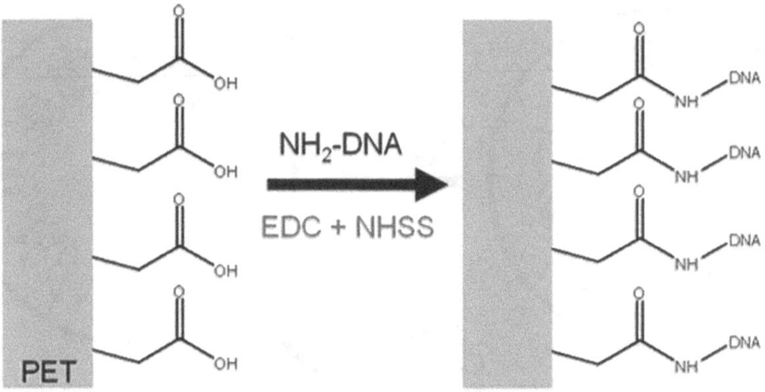

Fig. 2.1 Immobilization of G4 DNA onto the inner wall of the nanochannel by a two-step chemical reaction. Reprinted with the permission from Ref. [1]. Copyright 2009 American Chemical Society

allows control over the shape of the channels, and in this work the etched single nanochannel is cone-like. Diameter measurement of the single conical nanochannel was conducted with a commonly used electrochemical method [37, 39]. Its wide opening (base) is usually several hundred nanometers, and narrow opening (tip) is ~20 nm. During the etching process, negatively charged carboxyl groups, which were attached to flexible polymer chains, were created on the channel surface. Then, the amino single-stranded G4 DNA was immobilized onto the inner wall of the nanochannel by a two-step chemical reaction [8] (Fig. 2.1). After grafting G4 DNA on a conical nanochannel wall, the DNA underwent a potassium-responsive conformational change between a random single-stranded structure (without presence of K^+) and a four-stranded G4 structure (with presence of K^+), as shown in Fig. 2.2a, b. This conformational conversion could well induce a change in the effective channel size of the nanochannel, and thus realized K^+-responsive ion transport properties of the BNCS.

2.2 Materials and Methods

2.2.1 Materials

See Tables 2.1 and 2.2.

2.2.2 Instruments

See Table 2.3.

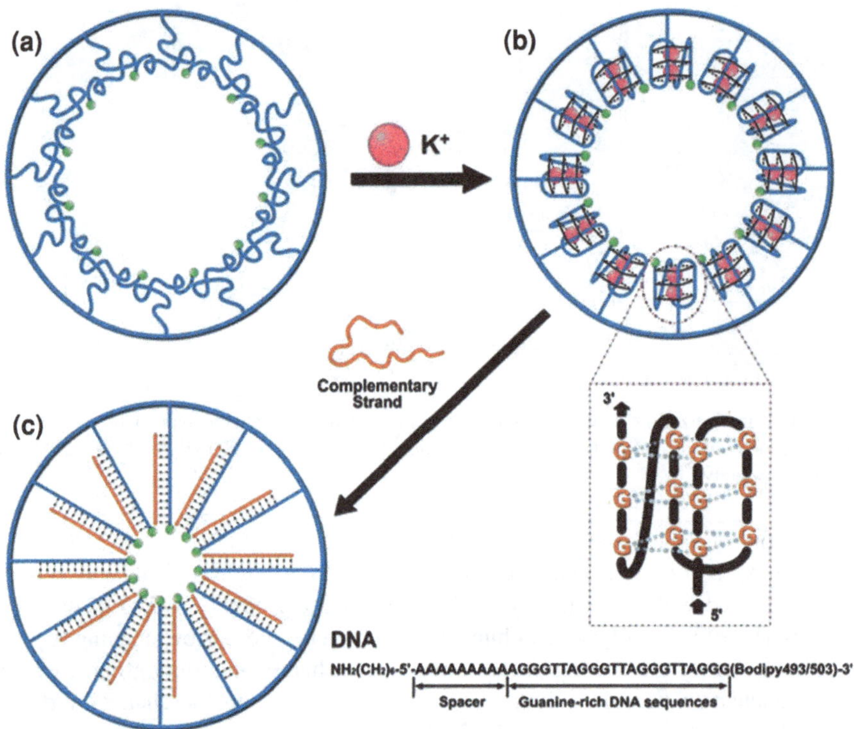

Fig. 2.2 G4 DNA was immobilized onto the inner surface of a single nanochannel. **a** There is no K^+; the G4 DNA relaxes to a loosely packed single-stranded structure. **b** Presence of K^+; the G4 DNA folds into densely packed rigid quadruplex structures that partially decrease the effective channel size of the nanochannel. **c** After adding complementary DNA strands, G4 DNA forms a closely packed arrangement of double-stranded DNA on the single nanochannel. Before modification, the etched funnel-shaped single nanochannels are around 20 nm wide at the narrowest point. Reprinted with the permission from Ref. [1]. Copyright 2009 American Chemical Society

Table 2.1 Samples used in the experiments

Sample	Purity and related parameters	Provider
Polyethylene terephthalate (PET) membrane	12 μm thick, with single ion track in the center	GSI, Germany
PET membrane	12 μm thick, with multi-ions track in the center (10^7/cm^2)	
PET membrane	12 μm thick, without any treatment	Hostaphan RN12 Hoechst, Germany
DNA probe (our design)	HPLC	TAKARA Biotechnology (DaLian) Co., Ltd., China
Primed DNA (our design)	PAGE	

Table 2.2 Reagents used in the experiments

Reagent	Purity quotient	Provider
KCl	A.R.	Beijing Chemical Works
LiCl	A.R.	Beijing Yili Fine Chemical Co., Ltd.
HCl	A.R.	Beijing Chemical Works (Contents 36–38 %)
Tris	A.R.	Made in China
NaOH	A.R.	Beijing Chemical Works
HCOOH	A.R.	Beijing Chemical Works (Contents >88 %)
EDC	≥99.0 % AT	Fluka
NHSS	≥98.5 % HPLC	Fluka
Filamentary silver	99.99 %	Made in China
Platinum filament	99.99 %	Made in China

Table 2.3 Instruments used in the experiments

Instrument	Producer
Water purification system (water, 18.2 MΩ)	Milli-Q, Millipore Corporation, USA
KEITHLEY 6487 picoammeter	Keithley Instruments, Cleveland, OH, USA
ZF-7 Uviol lamp	Shanghai Gucun Optic Instrument Factory, China
Model PHS-25 pH meter instruction	Shanghai Precision & Scientific Instrument Co. Ltd
Jasco J-810 circular dichroism (CD) spectropolarimeter	JASCO Corporation, Japan
Field emission scanning electron microscope, JSM-6700F	JEOL, Japan
X-ray photoelectron spectroscopy, ESCALab220i-XL	Thermo VG Scientific Ltd., UK
Homemade PTFE electrolyzer	ICCAS, China
Homemade plexiglass electrolyzer	ICCAS, China
Hot plate clarkson H3400-HS07	IKA, USA
Stopwatch timer	Made in China
Electronic thermometre	Made in China

2.2.3 Solvent Preparation

Track etching solution for the nanochannel preparation: 9 M NaOH
Stopping solution for the etching solution: 1 M KCl + 1 M HCOOH
CD test solution: Tris (5 mM) + HCl (4.5 mM) + DNA (1 μM)
Transmembrane current test solution:
Tris (5 mM) + HCl (4.5 mM) + KCl or LiCl (0–1500 μM).

2.3 Experiment Operation

2.3.1 Ag/AgCl Electrode Preparation

The characteristics of the silver/silver chloride (Ag/AgCl) electrode:

1. Reversibility: According to the electrochemical reaction, Ag/AgCl electrode could conduct in two directions.
2. Consuming: After the prolonged current conduction, especially in the high current case, the cathode of AgCl will be consumed, therefore it should be regularly plated.
3. Cl^- dependence: In order to achieve good conductivity, Ag/AgCl electrode must be immersed in a solution containing Cl^- (at least 10 mM).

Both a clean platinum wire and a clean silver wire (after the surface cleaning process, the platinum wire electrode is used as a cathode; after sanding smooth process, the silver electrode is used as an anode) are inserted into the 0.1 M HCl solution, and the electroplating by the external direct current power supply can be controlled by the adjustable resistance. Controlling the current density about 5 mA cm^{-2} for 5 min (at the same current density, there is a positive correlation between the plating time and the AgCl coating thickness. Generally, lower current density and longer plating time will produce the AgCl electrode with better quality), the surface of the silver wire as the anode is coated with AgCl (grayish black). Next, the AgCl electrode can be connected by a conducting wire with the insulating package. AgCl electrode is difficult to preserve, due to the photodecomposition. In addition, the AgCl coating will fall off when it dries. Therefore, after the experiment, the AgCl electrode needs to be washed with the deionized water and then is immersed in KCl solution for next use.

2.3.2 Ion Transport Measurement System

As shown in Fig. 2.3, the homemade transmembrane ion transport measurement equipment is a two-electrode system. The platinum electrode is used in the preparation of nanochannels, and the Ag/AgCl electrode is used for the ion transport experiments. Homemade polytetrafluoroethylene (PTFE) electrolyzer is mainly used in the preparation of nanochannels, because PTFE could be well suited for strong acid and alkali (Fig. 2.3a). PTFE three-cell system is used for the parallel etching experiments to get the single channel and multiple channels at the same time (Fig. 2.3b). PTFE two-cell system is individually used in the preparation of the single channel or multiple channels. Homemade plexiglass electrolyzer is used for the ion transport experiments, the advantage is that it can be well observed whether there is a generation of bubbles during the test, which will affect tests. Figure 2.3c shows the photo of ion transport measurement system. The red line is the output of the positive electrode, and the black line is the negative electrode. The electrode for the output voltage is also a test current electrode. The operating system of the instrument is Keithley Instruments ExceLINX software for the Model 6487, and test voltage range is between −10 and +10 V. Keithley 6487 can be used to conduct a variety of testings, such as potentiostatic tests and triangle scanning field tests.

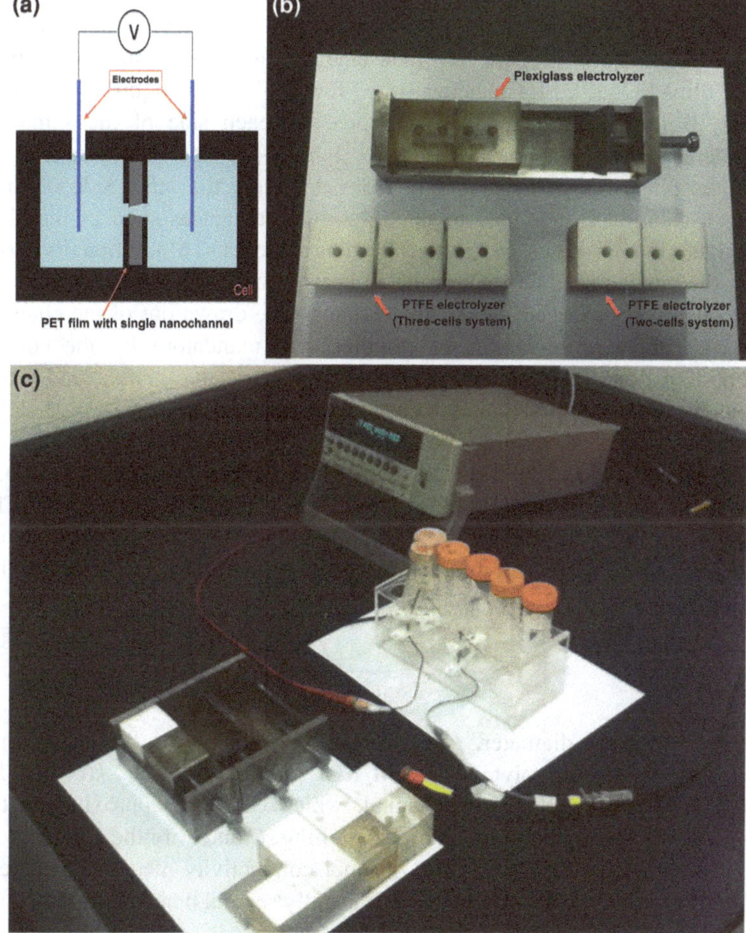

Fig. 2.3 **a** Schematic image of PET film with a conical nanochannel and the cell used for etching and to make all current measurements. **b** The cells. **c** The current measurement apparatus

2.3.3 Preparation of the Conical Shaped Single Nanochannel

The single conical nanochannel investigated here was produced in polymer films using the ion-track-etching technique. This method, which has become well established to create very uniform channel in insulator, is based on the following process (Fig. 1.6): when a swift heavy ion passes through the film, it deposits its energy along its trajectory, thus creating a cylindrical damage zone, i.e., the latent track. By the suitable wet etchants, the damaged material along the track can be removed more quickly than the bulk material, thus developing the tracks into nanochannels. Under the suitable conditions, channels down to a few nanometers in diameter can be produced. During the etching process, negatively charged carboxyl groups, which were attached to flexible polymer chains, were created on the channel surface.

Polyethylene terephthalate (PET, Hostaphan RN 12, Hoechst, thickness 12 μm) is a good candidate to prepare ion track channels with high aspect ratio in free-standing films [37–39]. PET is better suited for applications requiring very narrow channels. Before the chemical etching process, each side of the sample was independently exposed to the UV light for 1 h [36]. To produce a conical nano-channel, etching was performed only from one side, the other side of the cell contains a solution that is able to neutralize the etchant as soon as the channel opens, thus slowing down the further etching process. To additionally stop the etching, the voltage used to monitor the etching process was applied in such a way that the negatively charged ions of the etchant were drawn out of the channel tip (Fig. 1.6). This twofold automatic stopping was mandatory for the controlled production of nanosized channels. The following are the etching and stopping solutions for the etching of PET: 9 M NaOH for etching, 1 M KCl + 1 M HCOOH for stopping. The large opening of the conical nanochannel was called base, while the small opening was called tip. The diameter of the base was estimated from the bulk etch rate measured in the parallel etching experiments (Figs. 2.4, 2.5). The tip diameter was evaluated by the current measurement of the ion conductance of the nanochannel filled with 1 M potassium chloride solution as electrolyte via the following equation:

$$d_{\text{tip}} = \frac{4LI}{\pi k(c)UD} \quad (1)$$

where d_{tip} is the tip diameter, D is the base diameter, and $k(c)$ is the special conductivity of the electrolyte. For 1 M KCl solution at 25 °C, $k(c)$ is 0.11173 $\Omega^{-1}\text{cm}^{-1}$. L is the length of the channel, which could be approximated to the thickness of the membrane after chemical etching. U and I are the applied voltage and the measured ion current in the channel conductivity measurement, respectively. In this work, the base diameter was usually several hundred nanometers and the tip diameter was around 20 nm.

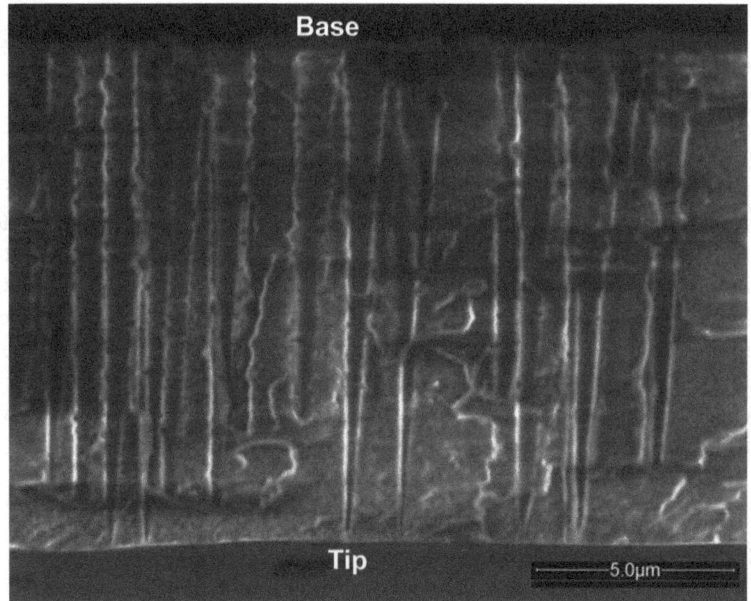

Fig. 2.4 The SEM of the profile of nanochannels, X8000. Scale bar, 5 μm

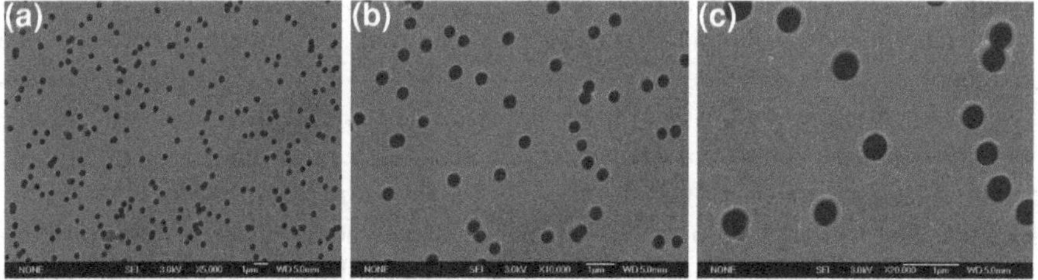

Fig. 2.5 The SEM of base sizes of nanochannels. **a** X5000. **b** X10000. **c** X20000. Scale bars, 1 μm

2.3.4 CD Spectroscopy Measurements

CD spectra were collected on a JASCO J-810 CD spectrometer. CD spectra were measured at 23 °C, as maintained by the temperature-control units affiliated to the spectrometers. Wavelength scans were performed between 220 and 320 nm. Quartz cells with a path length of 1 mm were used for DNA and tris solutions. The DNA was dissolved in a Tris buffer solution (pH = 7.2) containing Tris (5 mM) and HCl (4.5 mM) to give the DNA-Tris buffer solution a final concentration of 1 μM in a quartz cell. CD-melting profile recorded at 290 nm. The temperature increased from 10 to 90 °C. The average heating rate was about 1 °C/min.

2.3.5 DNA Immobilization

The amino single-stranded G4 DNA (5′-(NH_2)-(CH_2)$_6$-AAA AAA AAA AGG GTT AGG GTT AGG GTT AGG G (Bodipy493/503)-3′) on 5′ end (Fig. 2.2) and poly-A DNA(5′-(NH_2)-(CH_2)$_6$-AAA AAA AAA AAA AAA AAA AAA AAA AAA AAA A (Bodipy493/503)-3′) on 5′ end were immobilized onto the PET surface and inner channel wall by a two-step chemical reaction, as illustrated in Fig. 2.1. The NHSS ester was formed by exposure of the single-channel contained PET film to an aqueous solution (1 ml MilliQ water, 18.2 MΩ) of 15 mg EDC and 3 mg NHSS for 1 h at 23 °C. These PET-NHSS ester monolayers were reacted for 2 h with a solution of 1 μM amino DNA in MilliQ water at 23 °C. Then, the PET film had been stored for one day in Tris buffer (pH = 7.2, 23 °C) containing Tris (5 mM) and HCl (4.5 mM) before further experiments. The chemical covalent modification in this system is irreversible.

2.3.6 Current Measurement

The nanochannel conductivity and the ion responsive properties of the DNA molecules were studied by measuring ion current through the unmodified nanochannels or DNA-modified nanochannels. Ion current was measured by a Keithley 6487 picoammeter. A single-channel PET membrane was mounted between two chambers of the etching cell mentioned above (Fig. 2.3a). Ag/AgCl electrodes were used to apply a transmembrane potential across the film. Forward voltage was the potential applied on the base side (Fig. 1.6). The main transmembrane potential used in this work was a scanning voltage varied from −2 to +2 V with a 40 s period. Current measurement on the sample treated with Tris solution containing different concentration of alkali metal ions for 1 h before data collection at 23 °C. Each test was repeated five times to obtain the average current value at different voltage.

The previous work has discussed the relationship between positive and negative voltage for the transmembrane potential [8]. It was noticed that ion current signals of the conical nanochannel at low voltages were difficult to distinguish the change of ion transport properties under the little change of ion concentration of the electrolyte solution. Moreover, the electrode might be destroyed at high voltage. Therefore, in this work, the effect of ionic transport properties with different K^+ concentration on the BNCS is studied at a specific potential in one direction (2 V, anode facing the tip of nanochannel).

2.3.7 XPS Testing

X-ray photoelectron spectroscopy (XPS) analysis was utilized to detect the content of fluorine contained in the fluorescent group at 3′ end of the attached DNA molecule, in order to confirm that the DNA was successfully immobilized on PET surface, which was treated by Tris solution after immobilization (pH = 7.2, 296 K). XPS data were obtained by an ESCALab220i-XL electron spectrometer from VG Scientific using 300 W Al Kα radiation, and the base pressure was about 3×10^{-9} mbar. The binding energies were referenced to the C1s line at 284.8 eV from adventitious carbon. The XPS test was performed on a separate DNA attached PET surface, which was pretreated by 9 M sodium hydroxide before grafting DNA.

2.4 Results and Discussion

The conformational change of G4 DNA was determined by CD spectroscopy measurements. Control experiments showed that Li^+ had no obvious CD signal within the measured range when the concentration of Li^+ increased from 0 to 10 mM (Fig. 2.6a). As shown in Fig. 2.6b, when the concentration of K^+ increased from 0 to 10 mM, G4 DNA showed a positive peak near 290 nm, a crossover at around 260 nm, and weak negative peaks near 255 and 235 nm, thereby indicating a typical G4 conformation [22, 40]. According to the positive peak near 290 nm, when the concentration of K^+ increased up to 100 mM, the CD signal did not change significantly. It indicated that the major structure of this DNA was G4 conformation within the concentration range from 0 to 10 mM. Considering that the BNCS was realized by folding and unfolding of the G4 DNA, K^+ concentration

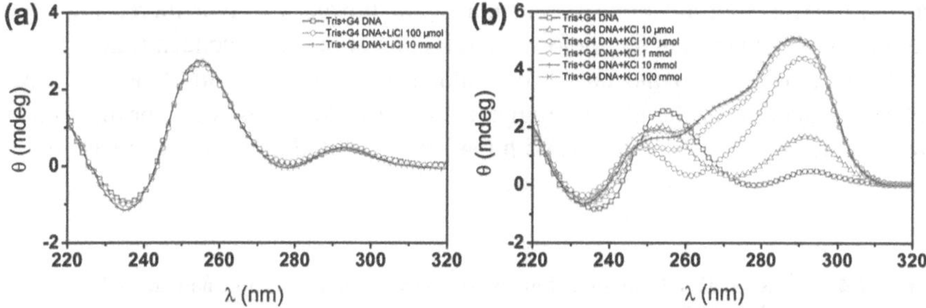

Fig. 2.6 CD spectra of G4 DNA (5′-GGG TTA GGG TTA GGG TTA GGG-3′) conformations in the different concentration of alkali metal ions. **a** The concentration of Li^+ of 0 μM (□, *black*), 100 μM (O, *red*), 10 mM (+, *blue*). **b** The concentration of K^+ of 0 μM (□, *black*), 10 μM (Δ, *green*), 100 μM (O, *red*), 1 mM (◇, *cyan*), 10 mM (+, *blue*), 100 mM (X, *magenta*) in Tris-HCl (5 mM, pH 7.2 at 23 °C). Reprinted with the permission from Ref. [1]. Copyright 2009 American Chemical Society

range was determined by G4 conformational change. Thus, this K^+ concentration range was considered as the maximum range for the BNCS current measurements.

In addition, owing to the molecules of fluorin contained in the fluorescent group Bodipy493/503 (green circle, Fig. 2.2) at 3′ end of the attached DNA molecule, DNA-modified surface of the PET films were also determined by the X-ray photoelectron spectroscopy analysis. Table 2.4 shows the PET surface without any DNA modification, and PET surface was treated by Tris solution at the same conditions. As the control experiments, it did not contain F1 s signal. Tables 2.5 and 2.6 show the fluorescent groups (Binding Energy of F1 s), indicating the PET surface attached with G4 DNA and poly-A DNA. The PET surface modified by G4 DNA or poly-A DNA did not show obvious difference at F1s signal.

Ion transport properties of the nanochannel were examined by current measurements. Some previous studies related to PET nanochannels have shown that there were two major influencing factors bringing the change of measuring current at the same voltage: ion concentration of the electrolyte solution [38, 41–47] and the effective channel size of the nanochannel [8]. It was assumed that the effective channel size of the nanochannel was determined by two aspects in this BNCS, i.e., the change of physical block and charge density caused by the biomolecular conformational change. In our strategies, the physical block and charge density would change simultaneously because the responsive DNA molecules were highly negatively charged in the pH neutral environment. In the following experiments, these two factors were integrated to investigate the 'effective channel size' change, which could more comprehensibly explain the new phenomenon in this BNCS. According to the fact that the nanochannel wall [9–11, 48] before and after G4 DNA modification is negatively charged under the neutral conditions, the electrolyte solutions in all the experiments were buffered to a pH value of 7.2 using 5 mM Tris-HCl at 23 °C in order to avoid the influence of pH.

Figure 2.7a shows a positive correlation between Li^+ concentration and the current before G4 DNA modification, which means that the current increased with Li^+ concentration from 0 to 1,500 μM at 2 V (anode facing the tip of nanochannel). After modification, the currents also showed a similar increasing trend but they were lower than that before modification at the same concentration, which could be attributed to the covalently attached DNA molecules that induced a relative reduction of the effective channel size [6]. Whereas the hybridization of the complementary DNA strands with G4 DNA resulted in a sharp decrease in the currents, it still kept an increasing trend.

Table 2.4 The XPS data from PET film before DNA immobilization and treated with Tris solution at pH 7.2 at 23 °C

Name (eV)	Start BE	Peak BE	End BE	Height counts	FWHM eV	Area (P) CPS.eV	At. %
C1s, 284.8	291.89	284.8	281.2	19890.06	1.89	51689.91	78.11
O1s, 532.2	537.03	532.33	527.9	10597.05	2.97	32997.46	20.88
N1s, 399.5	404.29	399.52	396.89	429.48	0.73	1093.24	1.01

Reprinted with the permission from Ref. [1]. Copyright 2009 American Chemical Society

Table 2.5 The XPS data from PET film after G4 DNA immobilization and treated with Tris solution at pH 7.2 at 23 °C

Name (eV)	Start BE	Peak BE	End BE	Height counts	FWHM eV	Area (P) CPS.eV	At. %
C1s, 284.8	293.2	284.81	281.34	13131.78	1.74	34589.26	73.73
O1s, 531.9	536.66	531.85	527.6	8545.3	1.91	27173.69	24.25
N1s, 399.7	405.49	399.68	395.37	518.66	1.36	1314.36	1.71
F1s, 689. 1	692.55	689	684.69	157.11	0	442.14	0.31

Reprinted with the permission from Ref. [1]. Copyright 2009 American Chemical Society

Table 2.6 The XPS data from PET film after poly-A DNA immobilization and treated with Tris solution at pH 7.2 at 23 °C

Name (eV)	Start BE	Peak BE	End BE	Height counts	FWHM eV	Area (P) CPS.eV	At. %
C1s, 284.8	293.26	284.79	281.14	10023.19	1.81	27783.84	70.06
O1s, 532.2	536.72	531.93	527.65	8329.17	3.01	26007.9	27.46
N1s, 400.1	405.05	399.89	396.03	364.1	0.09	1263.26	1.95
F1s, 688.8	692.05	688.6	685.13	253.15	0.27	653.34	0.53

Reprinted with the permission from Ref. [1]. Copyright 2009 American Chemical Society

Figure 2.7b shows the current change of the nanochannel via the concentration at different states upon the addition of K^+. Before G4 DNA modification and after the hybridization, the currents indicated an increasing trend similar to that in Fig. 2.7a. It is worth specially mentioning that there was a remarkable difference after DNA modification. The currents first started to drop with K^+ concentration increasing from 0 to 500 μM, and then the currents changed a little at the concentration range from 500 to 750 μM. Afterward, the concentration further increased from 750 to 1,500 μM, and the currents showed an increasing trend again. This unusual phenomenon could be attributed to the formation of G4 structures that induced the relatively dense packing of DNA molecules on the inner wall of the nanochannel, resulting in an efficient decrease of the effective channel size and thus the current change. These results coincided with the previous studies on the four-stranded i-motif structure [8]. Before the formation of G4 structures, G4 DNA with the single-stranded structure loosely packing on the nanochannel wall could not efficiently reduce the effective channel size, leading to the currents increasing with Li^+ concentration changing from 0 to 1,500 μM (Fig. 2.7a). However, the hybridization of the complementary DNA strands with G4 DNA formed the rigid duplex structure of DNA, and thus created a closely packed arrangement of double-stranded DNA structure that was more stable than G4 structure (Fig. 2.2c). Therefore, G4 DNA conformation could not change with K^+ concentration increasing. This deduction was strongly supported by further experimental data from the CD method.

As shown in Fig. 2.8, after adding the complementary DNA strands, the rigid duplex structures of DNA formed. No change could be observed in CD spectra when K^+ concentration changed from 0 to 1,500 μM, which meant the double-stranded DNA conformation was more stable than that at other states.

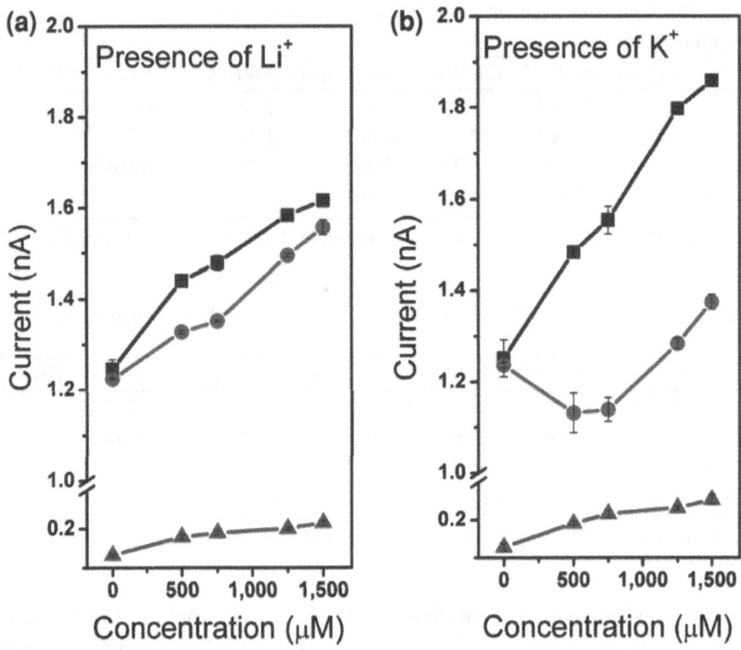

Fig. 2.7 Current-concentration (*I-C*) properties of a single nanochannel embedded in a PET membrane before and after G4 DNA molecules attached onto the channel wall in Tris-HCl (5 mM, pH = 7.2, at 23 °C). **a** Presence of different Li^+ concentration. **b** Presence of different K^+ concentration. Before G4 DNA modification (■, *blue*); after G4 DNA modification (●, *red*); after the addition of the complementary DNA strands (5′-CCC TAA CCC TAA CCC TAA CCC-3′) (▲, *green*). Before modification, the diameter of the tip and base are about 18 and 420 nm, respectively. Reprinted with the permission from Ref. [1]. Copyright 2009 American Chemical Society

To exclude the possible effect of regular DNA of which conformation was not responsive to K^+, poly-A DNA [8] was selected in this study. From Fig. 2.9, when the system was at its initial state without the alkali metal ions, poly-A DNA showed a negative peak near 247 nm. No change could be observed in CD spectra when the concentration of both K^+ and Li^+ of the buffer solution changed from 0 to 10 mM (G4 conformation appeared at the same K^+ concentration), which indicated there was no distinct conformational change, strongly suggesting that the conformation of poly-A DNA was not responsive to K^+ or Li^+.

Figure 2.10 shows the current change of a conical nanochannel upon the addition of Li^+ or K^+ before and after poly-A DNA modification under the same conditions, respectively. The ion transport properties of the nanochannel at different states also showed the increasing trend. It was reasonable that poly-A DNA with single-stranded structure loosely packing on the channel wall could not efficiently reduce the effective channel size, leading to the current increase at the ion concentration ranging from 0 to 1,500 μM. These results again verified that the

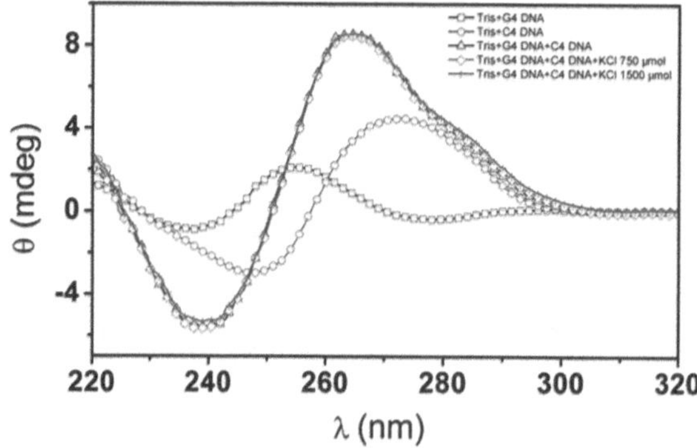

Fig. 2.8 CD spectra of G4 DNA (5′-GGG TTA GGG TTA GGG TTA GGG-3′), the complementary DNA strands (5′-CCC TAA CCC TAA CCC TAA CCC-3′) and double-stranded DNA (G4 DNA and the complementary DNA strands). G4 DNA in Tris-HCl, (□, *black*); the complementary DNA strands in Tris-HCl, (○, *red*); double-stranded DNA in the different concentration of K^+ ions, (0 μM, Δ, *blue*); (750 μM, ◇, *green*); (1500 μM, +, *magenta*). Reprinted with the permission from Ref. [1]. Copyright 2009 American Chemical Society

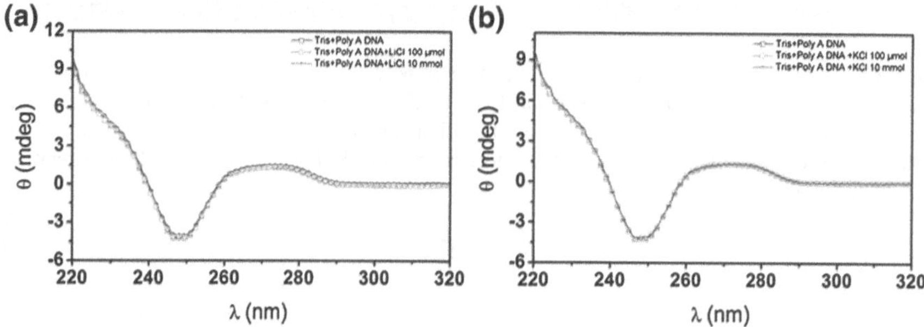

Fig. 2.9 CD spectra of poly-A DNA (5′-AA AAA AAA AAA AAA AAA AAA-3′) conformations in the different concentration of alkali metal ions. **a** Li^+. **b** K^+. 0 μM (□, *black*), 100 μM (○, *red*), 10 mM (+, *blue*) in Tris-HCl (5 mM, pH 7.2 at 23 °C). Reprinted with the permission from Ref. [1]. Copyright 2009 American Chemical Society

new phenomenon of the current change for K^+ after G4 DNA modification was mainly due to the conformation change of G4 DNA.

In this work, the diameter of the tip around 20 nm was selected to study this BNCS as the optimized channel size discussed in the previous work [8]. As shown in Fig. 2.11, all the nanochannels before modification, of which the tip diameters range from 17 to 24 nm, had the current ratios higher than 1, and exhibited the synthetic nanochannel original property that the current measurements had a positive correlation with alkali metal ion concentration. After the nanochannels were modified with G4 DNA, their current ratios showed the same trend for Li^+, but a totally different trend for K^+, which had the current ratios less than 1. Compared to the original nanochannel, the nanochannels modified with poly-A

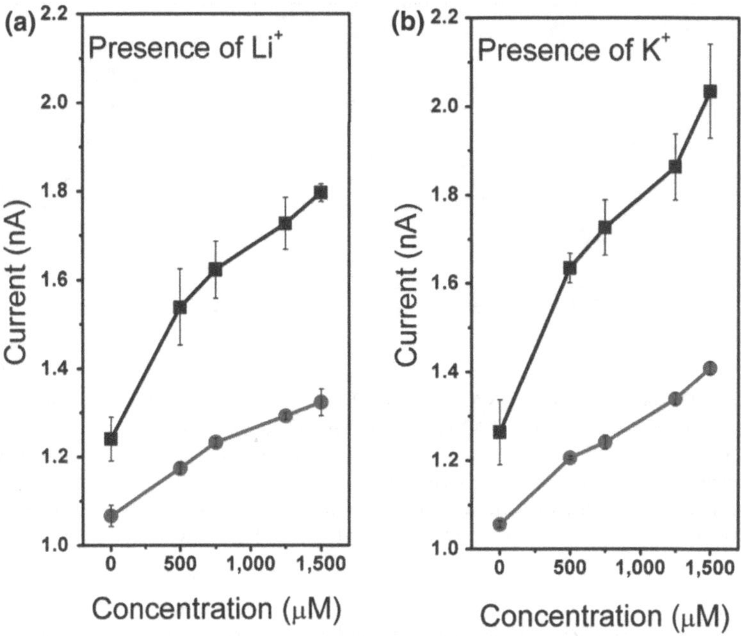

Fig. 2.10 *I–C* properties of a single nanochannel before and after poly-A DNA molecules attached onto the inner channel wall. **a** Li^+. **b** K^+. Before modification (■, *blue*); after modification (●, *red*). Before modification, the diameter of the tip and base are about 17 and 400 nm, respectively. Reprinted with the permission from Ref. [1]. Copyright 2009 American Chemical Society

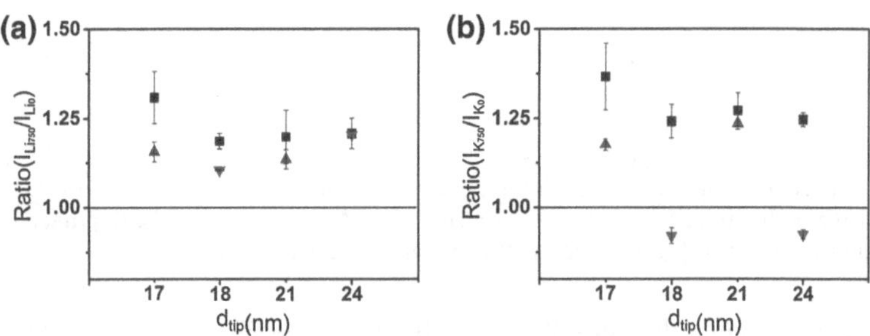

Fig. 2.11 Current ratio of the nanochannels under the absence and presence of 750 μM alkali metal ions before and after DNA modification. **a** Li^+, **b** K^+. Before modification (■, *blue*), after G4 DNA modification (▼, *red*), after poly-A DNA modification (▲, *green*). The diameters of the tip and base are about 20 and 400 nm (Sample1, ~17, ~400 nm; Sample2, ~18, 420 nm; Sample3, ~21, ~400 nm; Sample4, ~24, ~410 nm). Reprinted with the permission from Ref. [1]. Copyright 2009 American Chemical Society

DNA showed the similar increasing trend for both Li^+ and K^+. Therefore, the conformational change exhibited in G4 DNA molecules indeed contributes to this new phenomenon in the BNCS.

To explore the phenomenon observed in Fig. 2.7b that the currents first decreased and then increased with changes of K^+ concentration, the CD melting method was utilized [21, 23]. It is well-known that G4 DNA melting reveals the ratio of folding and unfolding of G4 DNA, and corresponds to the stability of the G4 structures [21]. Figure 2.12 indicated that the stability of the G4 structures increased with K^+ concentration increasing. The cyan horizontal dot line represents the temperature (T_{et}) for the current measurement. When K^+ concentration exceeded C_{et} (the corresponding concentration of the intersection point between the black curve and the cyan horizontal dot line), the G4 structures became more and more stable with K^+ concentration increasing. With the stability of structures further enhanced, however, the change of the DNA melting rate showed a type of "L" curve (Fig. 2.12, inserts), which means that, for this kind of G4 DNA, K^+ concentration reached a certain level, the stability of the G4 structures nearly kept unchanged. It was found that the most prominent stabilizing of the G4 DNA structure appeared below K^+ concentration of 750 μM.

Based on the above analyses, it can be concluded that enhancing the stability of the G4 structures by increasing K^+ concentration could gradually decrease the effective channel size of the nanochannel within a certain ion concentration ranging from 0 to 750 μM. On the other aspect, K^+ concentration exhibited a positive correlation with the currents change (Fig. 2.7b). Therefore, this new phenomenon of the ion transport properties of the nanochannel modified with G4 DNA was explained by the competition between these two factors, i.e., the stability of G4 structures and ion concentration. To be clearer, when the stability of G4 structures dramatically increased with K^+ concentration increasing, it played a prominent role in changing ion transport properties of the nanochannel by adjusting the effective channel size, while when the increase rate of the stability of G4 structures gradually decreased, the ion concentration started to exert main influence.

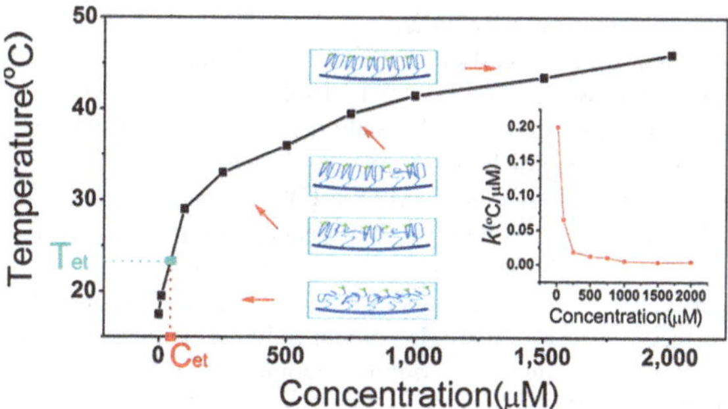

Fig. 2.12 DNA melting—Ion concentration curve of G4 DNA (5′-GGG TTA GGG TTA GGG TTA GGG-3′). The *inserts* represent the stability of G4 structure under different K^+ concentration at the fixed experimental temperature, and influence of different K^+ concentration on the slope of G4 DNA melting and K^+ concentration. Reprinted with the permission from Ref. [1]. Copyright 2009 American Chemical Society

2.5 Conclusions

In summary, we experimentally demonstrate a novel biomimetic nanochannel which can achieve a K^+ response within a certain ion concentration range. The situation of the grafting G4 DNA on a single nanochannel can closely imitate the in vivo condition [21], because the G-rich telomere overhang is attached to the chromosome. Therefore, this artificial system could promote a potential to conveniently study biomolecule conformational change in confined space by the current measurement which is significantly different from the nanopore sequencing [49]. Moreover, such a system as a basic platform could potentially spark further experimental and theoretical efforts to simulate the process of ion transport in living organism and boost the development of bio-inspired intelligent nanochannel apparatus such as biosensors [50, 51], nanofluidic devices [52] and molecular filtration [53].

References

1. Hou X, Guo W, Xia F, Nie FQ, Dong H, Tian Y, Wen LP, Wang L, Cao LX, Yang Y, Xue JM, Song YL, Wang YG, Liu DS, Jiang L (2009) A biomimetic potassium responsive nanochannel: G-Quadruplex DNA conformational switching in a synthetic nanopore. J Am Chem Soc 131(22):7800–7805. doi:10.1021/Ja901574c
2. Apel P (2001) Track etching technique in membrane technology. Radiat Meas 34(1–6):559–566. doi:10.1016/s1350-4487(01)00228-1
3. Harrell CC, Kohli P, Siwy Z, Martin CR (2004) DNA—Nanotube artificial ion channels. J Am Chem Soc 126(48):15646–15647. doi:10.1021/ja044948v
4. Kohli P, Harrell CC, Cao ZH, Gasparac R, Tan WH, Martin CR (2004) DNA-functionalized nanotube membranes with single-base mismatch selectivity. Science 305(5686):984–986. doi:10.1126/science.1100024
5. Schmuhl R, van den Berg A, Blank DHA, ten Elshof JE (2006) Surfactant-modulated switching of molecular transport in nanometer-sized pores of membrane gates. Angew Chem Int Edit 45(20):3341–3345. doi:10.1002/anie.200504579
6. Nilsson J, Lee JRI, Ratto TV, Letant SE (2006) Localized functionalization of single nanopores. Adv Mater 18(4):427–431. doi:10.1002/adma.200501991
7. Iqbal SM, Akin D, Bashir R (2007) Solid-state nanopore channels with DNA selectivity. Nat Nanotechnol 2(4):243–248. doi:10.1038/nnano.2007.78
8. Xia F, Guo W, Mao YD, Hou X, Xue JM, Xia HW, Wang L, Song YL, Ji H, Qi OY, Wang YG, Jiang L (2008) Gating of single synthetic nanopores by proton-driven DNA molecular motors. J Am Chem Soc 130(26):8345–8350. doi:10.1021/Ja800266p
9. Ali M, Yameen B, Neumann R, Ensinger W, Knoll W, Azzaroni O (2008) Biosensing and supramolecular bioconjugation in single conical polymer nanochannels. facile incorporation of biorecognition elements into nanoconfined geometries. J Am Chem Soc 130(48):16351–16357. doi:10.1021/ja8071258
10. Ali M, Ramirez P, Mafe S, Neumann R, Ensinger W (2009) A pH-tunable nanofluidic diode with a broad range of rectifying properties. ACS Nano 3(3):603–608. doi:10.1021/nn900039f
11. Yameen B, Ali M, Neumann R, Ensinger W, Knoll W, Azzaroni O (2009) Single conical nanopores displaying pH-tunable rectifying characteristics. manipulating ionic transport with zwitterionic polymer brushes. J Am Chem Soc 131(6):2070–2071. doi:10.1021/ja8086104

12. Parkinson GN, Lee MPH, Neidle S (2002) Crystal structure of parallel quadruplexes from human telomeric DNA. Nature 417(6891):876–880. doi:10.1038/nature755
13. Patel DJ (2002) Structural biology—a molecular propeller. Nature 417(6891):807–808. doi:10.1038/417807a
14. Davis JT (2004) G-quartets 40 years later: From 5'-GMP to molecular biology and supramolecular chemistry. Angew Chem Int Edit 43(6):668–698. doi:10.1002/anie.200300589
15. Dittmer WU, Reuter A, Simmel FC (2004) A DNA-based machine that can cyclically bind and release thrombin. Angew Chem Int Edit 43(27):3550–3553. doi:10.1002/anie.200353537
16. Alberti P, Bourdoncle A, Sacca B, Lacroix L, Mergny J-L (2006) DNA nanomachines and nanostructures involving quadruplexes. Org Biomol Chem 4(18):3383–3391. doi:10.1039/b605739j
17. Maizels N (2006) Dynamic roles for G4 DNA in the biology of eukaryotic cells. Nat Struct Mol Biol 13(12):1055–1059. doi:10.1038/nsmb1168
18. Monchaud D, Yang P, Lacroix L, Teulade-Fichou M-P, Mergny J-L (2008) A metal-mediated conformational switch controls G-quadruplex binding affinity. Angew Chem Int Edit 47(26):4858–4861. doi:10.1002/anie.200800468
19. Smargiasso N, Rosu F, Hsia W, Colson P, Baker ES, Bowers MT, De Pauw E, Gabelica V (2008) G-quadruplex DNA assemblies: loop length, cation identity, and multimer formation. J Am Chem Soc 130(31):10208–10216. doi:10.1021/ja801535e
20. Domene C, Klein ML, Branduardi D, Gervasio FL, Parrinello M (2008) Conformational changes and gating at the selectivity filter of potassium channels. J Am Chem Soc 130 (29):9474–9480. doi:10.1021/ja801792g
21. Zhao Y, Kan ZY, Zeng ZX, Hao YH, Chen H, Tan Z (2004) Determining the folding and unfolding rate constants of nucleic acids by biosensor. Application to telomere G-quadruplex. J Am Chem Soc 126 (41):13255–13264. doi:10.1021/ja048398c
22. Phan AT, Kuryavyi V, Luu KN, Patel DJ (2007) Structure of two intramolecular G-quadruplexes formed by natural human telomere sequences in K^+ solution. Nucleic Acids Res 35(19):6517–6525. doi:10.1093/nar/gkm706
23. Lane AN, Chaires JB, Gray RD, Trent JO (2008) Stability and kinetics of G-quadruplex structures. Nucleic Acids Res 36(17):5482–5515. doi:10.1093/nar/gkn517
24. Forman SL, Fettinger JC, Pieraccini S, Gottareli G, Davis JT (2000) Toward artificial ion channels: A lipophilic G-quadruplex. J Am Chem Soc 122(17):4060–4067. doi:10.1021/ja9925148
25. Kaucher MS, Harrell WA, Davis JT (2006) A unimolecular G-quadruplex that functions as a synthetic transmembrane Na^+ transporter. J Am Chem Soc 128(1):38–39. doi:10.1021/ja056888e
26. Sakai N, Kamikawa Y, Nishii M, Matsuoka T, Kato T, Matile S (2006) Dendritic folate rosettes as ion channels in lipid bilayers. J Am Chem Soc 128(7):2218–2219. doi:10.1021/ja058157k
27. Lee MPH, Parkinson GN, Hazel P, Neidle S (2007) Observation of the coexistence of sodium and calcium ions in a DNA G-quadruplex ion channel. J Am Chem Soc 129(33):10106–10107. doi:10.1021/ja0740869
28. Hennig A, Matile S (2008) Detection of the activity of ion channels and pores by circular dichroism spectroscopy: G-quartets as functional CD probes within chirogenic vesicles. Chirality 20(9):932–937. doi:10.1002/chir.20526
29. Ma L, Melegari M, Colombini M, Davis JT (2008) Large and stable transmembrane pores from guanosine-bile acid conjugates. J Am Chem Soc 130(10):2938–2939. doi:10.1021/ja7110702
30. Ueyama H, Takagi M, Takenaka S (2002) A novel potassium sensing in aqueous media with a synthetic oligonucleotide derivative. Fluorescence resonance energy transfer associated with guanine quartet-potassium ion complex formation. J Am Chem Soc 124 (48):14286–14287. doi:10.1021/ja026892f

31. He F, Tang YL, Wang S, Li YL, Zhu DB (2005) Fluorescent amplifying recognition for DNA G-quadruplex folding with a cationic conjugated polymer: a platform for homogeneous potassium detection. J Am Chem Soc 127(35):12343–12346. doi:10.1021/ja051507i
32. He F, Tang YL, Yu MH, Feng F, An LL, Sun H, Wang S, Li YL, Zhu DB, Bazan GC (2006) Quadruplex-to-duplex transition of G-rich oligonucleotides probed by cationic water-soluble conjugated polyelectrolytes. J Am Chem Soc 128(21):6764–6765. doi:10.1021/ja058075w
33. Huang C-C, Chang H-T (2008) Aptamer-based fluorescence sensor for rapid detection of potassium ions in urine. Chem Commun 12:1461–1463. doi:10.1039/b718752a
34. Phan AT, Mergny JL (2002) Human telomeric DNA: G-quadruplex, i-motif and watson-crick double helix. Nucleic Acids Res 30(21):4618–4625. doi:10.1093/nar/gkf597
35. Sen D, Gilbert W (1990) A sodium-potassium switch in the formation of 4-stranded G4-DNA. Nature 344(6265):410–414. doi:10.1038/344410a0
36. Apel PY, Korchev YE, Siwy Z, Spohr R, Yoshida M (2001) Diode-like single-ion track membrane prepared by electro-stopping. Nucl Instrum Meth Phys Res Sect B Beam Interact Mater Atoms 184(3):337–346. doi:10.1016/s0168-583x(01)00722-4
37. Siwy Z, Apel P, Baur D, Dobrev DD, Korchev YE, Neumann R, Spohr R, Trautmann C, Voss KO (2003) Preparation of synthetic nanopores with transport properties analogous to biological channels. Surf Sci 532:1061–1066. doi:10.1016/s0039-6028(03)00448-5
38. Harrell CC, Siwy ZS, Martin CR (2006) Conical nanopore membranes: Controlling the nanopore shape. Small 2(2):194–198. doi:10.1002/smll.200500196
39. Wharton JE, Jin P, Sexton LT, Horne LP, Sherrill SA, Mino WK, Martin CR (2007) A method for reproducibly preparing synthetic nanopores for resistive-pulse biosensors. Small 3 (8):1424–1430. doi:10.1002/smll.200700106
40. Xu Y, Noguchi Y, Sugiyama H (2006) The new models of the human telomere d AGGG (TTAGGG)(3) in K^+ solution. Bioorg Med Chem 14 (16):5584–5591. doi:10.1016/j.bmc.2006.04.033
41. Siwy Z, Fulinski A (2002) Fabrication of a synthetic nanopore ion pump. Phys Rev Lett 89 (19). doi:198103 10.1103/PhysRevLett.89.198103
42. Kumar S, Chakarvarti SK (2003) On the preparation and asymmetric electric transport behavior of conical channels in polyethylene terepthalate. Radiat Meas 36(1–6):757–760. doi:10.1016/s1350-4487(03)00241-5
43. Siwy Z, Apel P, Dobrev D, Neumann R, Spohr R, Trautmann C, Voss K (2003) Ion transport through asymmetric nanopores prepared by ion track etching. Nucl Instrum Methods Phys Res Sect B Beam Interact Mater Atoms 208:143–148. doi:10.1016/s0168-583x(03)00884-x
44. Schiedt B, Healy K, Morrison AP, Neumann R, Siwy Z (2005) Transport of ions and biomolecules through single asymmetric nanopores in polymer films. Nucl Instrum Meth Phys Res Sect B Beam Interact Mater Atoms 236:109–116. doi:10.1016/j.nimb.2005.03.265
45. Siwy Z, Kosinska ID, Fulinski A, Martin CR (2005) Asymmetric diffusion through synthetic nanopores. Phys Rev Lett 94(4). doi:10.1103/PhysRevLett.94.048102
46. Siwy ZS (2006) Ion-current rectification in nanopores and nanotubes with broken symmetry. Adv Funct Mater 16(6):735–746. doi:10.1002/adfm.200500471
47. Powell MR, Sullivan M, Vlassiouk I, Constantin D, Sudre O, Martens CC, Eisenberg RS, Siwy ZS (2008) Nanoprecipitation-assisted ion current oscillations. Nat Nanotechnol 3 (1):51–57. doi:10.1038/nnano.2007.420
48. Siwy Z, Heins E, Harrell CC, Kohli P, Martin CR (2004) Conical-nanotube ion-current rectifiers: the role of surface charge. J Am Chem Soc 126(35):10850–10851. doi:10.1021/ja047675c
49. Branton D, Deamer DW, Marziali A, Bayley H, Benner SA, Butler T, Di Ventra M, Garaj S, Hibbs A, Huang X, Jovanovich SB, Krstic PS, Lindsay S, Ling XS, Mastrangelo CH, Meller A, Oliver JS, Pershin YV, Ramsey JM, Riehn R, Soni GV, Tabard-Cossa V, Wanunu M, Wiggin M, Schloss JA (2008) The potential and challenges of nanopore sequencing. Nat Biotechnol 26(10):1146–1153. doi:10.1038/nbt.1495
50. Choi Y, Baker LA, Hillebrenner H, Martin CR (2006) Biosensing with conically shaped nanopores and nanotubes. Phys Chem Chem Phys 8(43):4976–4988. doi:10.1039/b607360c

51. Martin CR, Siwy ZS (2007) Learning nature's way: biosensing with synthetic nanopores. Science 317(5836):331–332. doi:10.1126/science.1146126
52. Huh D, Mills KL, Zhu X, Burns MA, Thouless MD, Takayama S (2007) Tuneable elastomeric nanochannels for nanofluidic manipulation. Nat Mater 6(6):424–428. doi:10.1038/nmat1907
53. Savariar EN, Krishnamoorthy K, Thayumanavan S (2008) Molecular discrimination inside polymer nanotubules. Nat Nanotechnol 3(2):112–117. doi:10.1038/nnano.2008.6

Chapter 3
Asymmetric pH-Gating Symmetric Hour-Glass Shaped Single Nanochannel

3.1 Introduction

Inspired by ion channels, the generation of artificial nanochannels has strong implications to simulate the different process of ionic transport as well as enhance the functionality of biological channels. Here I demonstrate a plasma asymmetric chemical modification approach to prepare the pH asymmetric gating nanochannel that can achieve pH control for both different ionic rectification and perfect gating function, simultaneously [1].

Ion channels that regulate ion permeation through cell membranes are important for the implementation of various significant physiological functions in life processes [2]. The components of these channels are asymmetrically distributed between membrane surfaces [3]. Inspired by these asymmetrical nanochannels including various components of ion channels which are not uniform in distribution and the structural asymmetry, the generation of artificial nanochannels has strong implications to simulate the different process of ion transport as well as enhance the functionality of biological ion channels [4]. Recently, Jiang and Azzaroni et al. have successfully developed simple-function pH-control nanochannels [5–8]. Here I further develop the concept of a smart nanochannel system, which is not subject to the solution environment restriction of the chemical modification [9], by using a plasma asymmetric chemical modification approach. Compared with other systems [5–8], this responsive nanochannel system has the advantage that it provides simultaneous control over the pH gating [8] (Fig. 3.1a) and pH-tunable asymmetric [5–7] (Fig. 3.1b) ionic transport properties. This highly effective method can be used in the near future to build smarter, biologically inspired nanochannel machines with more precisely controlled functions by designing more complicated functional molecules.

There has been rapid progress in developing chemical modification of the interior surface of nanochannels with various responsive properties, such as specific ions [10], light [11], pH [5–8], and temperature [12]. In order to achieve different functionalities of artificial nanochannels, various methods have been

X. Hou, *Bio-inspired Asymmetric Design and Building of Biomimetic Smart Single Nanochannels*, Springer Theses,
DOI: 10.1007/978-3-642-38050-1_3, © Springer-Verlag Berlin Heidelberg 2013

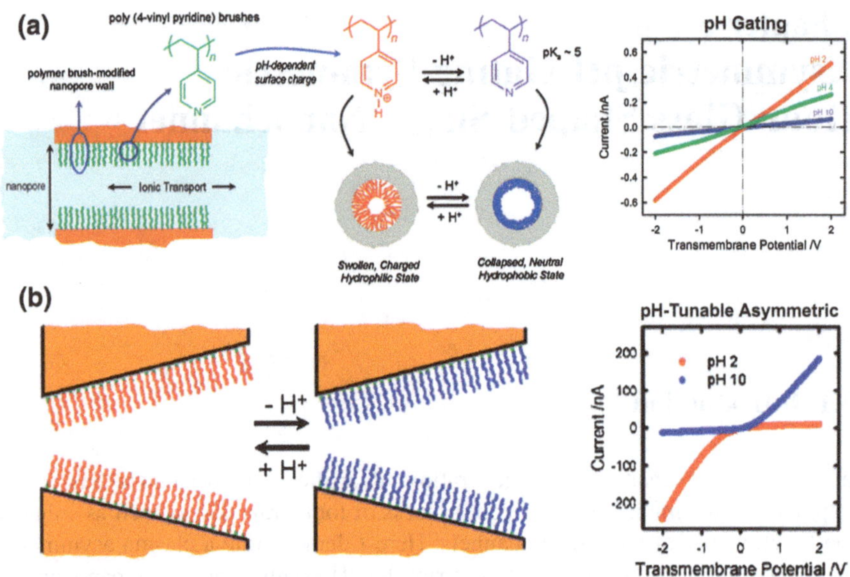

Fig. 3.1 *I–V* curves of symmetric (**a**) and asymmetric (**b**) single nanochannels after the functionalization. Reprinted with permission from Ref. [7, 8]. Copyright 2009 American Chemical Society

invented, such as electroless deposition [13], solution chemical modification [5], and electrostatic self-assembly [14], which can only cover the whole inner surface of the nanochannels.

Plasma technology, which mainly includes plasma etching and plasma modification, offers an effective method for nanoscale surface engineering of materials. Plasma etching has often been used in template synthesis of nanomaterials [15, 16]. Plasma modification can functionalize a specific local area precisely, whether via symmetric or asymmetric chemical modification. This advantage can provide the potential to achieve a variety of nanoscale features and new properties in developing advanced nanomaterials. Although multiple nanochannel membranes using plasma modification have been studied [17–19] for the control of the water permeability of polymeric membranes, we still need an optimal system, such as a single nanochannel system, for studying transport properties of different ionic or molecules in a confined space, without having to average the effects of multiple channels. Therefore, in this work, I developed a pH-gating ionic transport nanodevice using plasma asymmetric chemical modification, and this approach can be used as a platform for developing more ways to achieve the precise asymmetric chemical modification of the interior surface of nanochannels, which have strong implications for the simulation of different ionic transport processes as well as the enhancement of ion channel functionality.

3.2 Materials and Methods

3.2.1 Materials

See Tables 3.1 and 3.2.

Table 3.1 Samples used in the experiments

Sample	Purity and related parameters	Provider
PET membrane	12 μm thick, with single ion track in the center	GSI, Germany
PET membrane	12 μm thick, with multi-ions track in the center (10^7/cm^2)	
PET membrane	12 μm thick, without any treatment	Hostaphan RN12 Hoechst, Germany

Table 3.2 Reagents used in the experiments

Reagent	Purity quotient	Provider
KCl	A.R.	Beijing Chemical Works
HCl	A.R.	Beijing Chemical Works (Contents 36–38 %)
NaOH	A.R.	Beijing Chemical Works
KOH	A.R.	West Long Chemical Co., Ltd., China
HCOOH	A.R.	Beijing Chemical Works (Contents >88 %)
Acrylic acid (AAc)	A.R.	Aladdin, China
Filamentary silver	99.99 %	Made in China
Platinum filament	99.99 %	Made in China

3.2.2 Instruments

See Table 3.3.

3.2.3 Solvent Preparation

Track etching solution for nanochannel preparation: 9 M NaOH
Stopping solution for the etching solution: 1 M KCl + 1 M HCOOH
Transmembrane current test solution: KCl (0.1 and 1 M)

Table 3.3 Instruments used in the experiments

Instrument	Producer
Water purification system (water, 18.2 MΩ)	Milli-Q, Millipore Corporation, USA
KEITHLEY 6487 picoammeter	Keithley Instruments, Cleveland, OH, USA
OCA20, Contact angle (CA) measuring instrument	Dataphysics, Germany
DT-01 Plasma instrument	SuZhou OPS Plasma Technology Co., Ltd., China
DJ-01, Plasma graft instrument	SuZhou OPS Plasma Technology Co., Ltd., China
ZF-7 Uviol lamp	Shanghai Gucun Optic Instrument Factory, China
Model PHS-25 pH meter instruction	Shanghai Precision & Scientific Instrument Co., Ltd
Environmental Scanning Electronic Microscope Quanta 200 FEG	FEI Company, USA
Homemade PTFE electrolyzer	ICCAS, China
Homemade Plexiglass electrolyzer	ICCAS, China
Hot Plate Clarkson H3400-HS07	IKA, USA
Stopwatch timer	Made in China
Electronic thermometre	Made in China

3.3 Experiment Operation

3.3.1 Preparation of the Hour-Glass Shaped Single Nanochannel

The single hour-glass shaped nanochannel investigated here was produced in polymer films using the ion-track-etching technique [9, 10, 20, 21]. Before the chemical etching process, the samples of the PET membrane with single ion track in the center were exposed to UV light for 1 h for each side. To produce an hour-glass shaped nanochannel, etching was performed on both sides. For the observation of the etching process, the voltage (1 V) used to monitor the etching process was applied in such a way that the transmembrane ionic current can be observed as soon as the nanochannel opened. A solution was added to both sides of the cell to neutralize the etchant as soon as the nanochannel opened, and thus slowdown the further etching process. The following etching and stopping solutions are for the etching of PET: 9 M NaOH for etching, 1 M KCl + 1 M HCOOH for stopping. The opening of the hour-glass shaped nanochannel was called base, while the small center was called tip. The diameter of the base was estimated from the multitrack membranes etch rate measured in the parallel etching experiments by environmental scanning electronic microscope (ESEM, Fig. 3.2), due to the difficulty [22] of locating the base of the single nanochannel in ESEM. In this work, the base of the single channel was usually controlled from ~250 to ~300 nm, and its tip ranged from ~10 to ~30 nm.

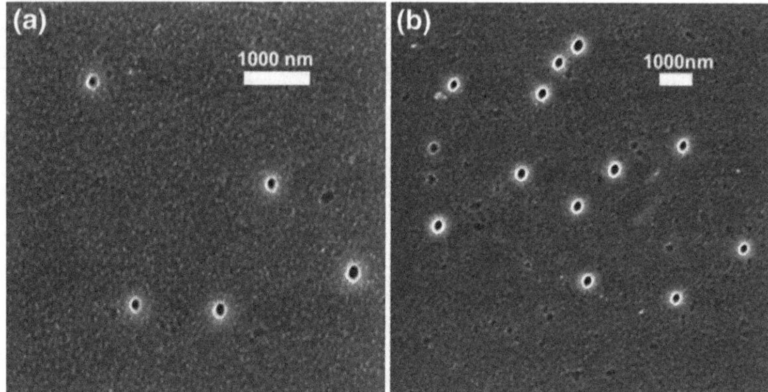

Fig. 3.2 The ESEM of channel sizes of multi-nanochannels on the different chemical etching times. **a** 10 min. **b** 15 min. *Scale bars*, 1000 nm. Reproduced from Ref. [1] by permission of John Wiley & Sons Ltd

3.3.2 Plasma-Induced Graft Polymerization

The PET film was first soaked in water for 5 h after the etching experiment. Then, distilled acrylic acid (AAc) was injected into a plasma-induced grafting reactor. Vacuum before switching on the glow discharge was 21 Pa and the working temperature was 20 °C. In my experiment setup, the argon atmosphere was kept at about ~50–70 Pa and these conditions should be kept for 15 min. Then, a Start R-F power supply source was applied at 20 W to obtain glow discharge, which was maintained for 30 min. After the glow extinguished, grafting of the AAc monomers would take place in the grafting reactor, where a vacuum of ~300–600 Pa was maintained for about 20 min. Finally, the chamber was connected with air and the plasma treatment was finished.

3.3.3 Current Measurement

The ionic transport properties of the nanochannel were studied by measuring the ionic current through the nanochannels before and after plasma treatment. Ionic current was measured by a Keithley 6487 picoammeter. A single hour-glass shaped PET membrane was mounted between two chambers of the etching cell mentioned above. Ag/AgCl electrodes were used to apply a transmembrane potential across the film. The forward voltage was the potential applied on one of the base sides, which was the opposite side of the plasma treatment. The main transmembrane potential used in this work was evaluated and the scanning voltage

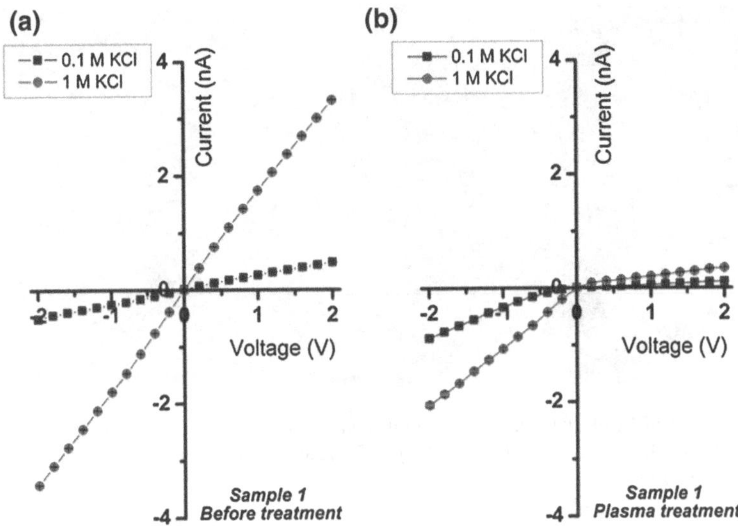

Fig. 3.3 The different concentrations of KCl solutions for the current measurements of the single nanochannel before and after plasma treatment. **a**, **b** the *I–V* properties of the nanochannels with the different concentrations (0.1 M, ■, *blue*; 1 M, ●, *red*;). (Sample1, base ~250 nm, tip ~11 nm, before plasma treatment). Reproduced from Ref. [1] by permission of John Wiley & Sons Ltd

varied from −2 to +2 V with a 40 s period. The electrolyte concentration was also evaluated (Fig. 3.3). Finally, 0.1 M potassium chloride solution was chosen as electrolyte. The pH of the electrolyte was adjusted with 1 M HCl and KOH solutions, and the influence of addition substance quality can be ignored. Each test was repeated 5 times to obtain an average current value at different voltages. The testing temperature was 20 °C.

3.3.4 CA Measurement

Contact angles (CA) were measured using an OCA20 contact-angle system at ambient temperature and saturated humidity. The original sample was treated with 9 M NaOH for 15 min. The sample was then taken out from the etching solution and treated with the stopping solution (1 M KCl + 1 M HCOOH) for half an hour. After that, the sample was treated with the deionized water, and had been stored for 5 h in the deionized water before further experiments. Before the CA test, the sample was blown dry by N_2. Deionized water droplets (about 2 μl) were then dropped carefully onto the surfaces. The average contact angel value was obtained at five different positions of the same sample. Finally, the results were summarized.

3.4 Results and Discussion

In this work the etched single nanochannel is symmetric and hour-glass shaped (Fig. 3.4). Diameter measurements of hour-glass shaped nanochannels were conducted with a commonly used electrochemical method [21]. The opening at the base was usually ~250–300 nm wide and the narrow center (tip) was ~10–30 nm wide. Only one side of the nanochannel was treated by plasma-induced grafting in the vapor phase of the distilled AAc, which became a pH-responsive polymer [23, 24], poly acrylic acid (PAA), after plasma-induced graft polymerization. To explore the surface properties of the PET films before and after plasma treatment, the films have been studied by CA measurements. The CA of the PET film surfaces after plasma treatment (Fig. 3.5b) was smaller than the one before treatment (Fig. 3.5a). The results of the CA measurements showed that the plasma treatment could lead to a visible change of the surface wettability (from $66.6^{\circ} \pm 1.3^{\circ}$ to $36.7^{\circ} \pm 6.9^{\circ}$), which indicated a change in the chemical composition. This result may spark further experimental approaches to study the relationship between the wettability and ionic transport properties within a confined space at the nanoscale [25].

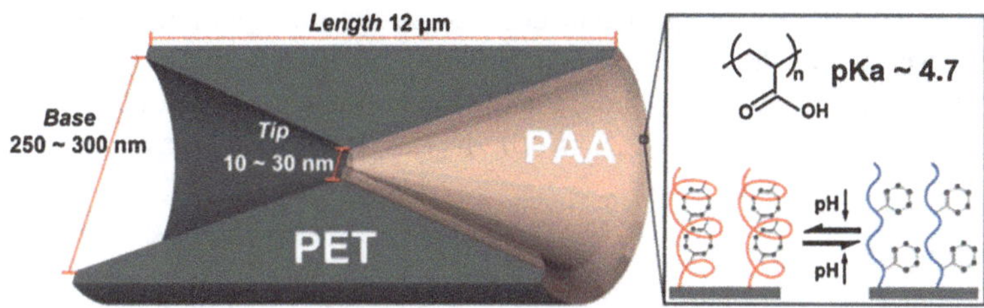

Fig. 3.4 Scheme of the single hour-glass shaped nanochannel after plasma-induced graft polymerization (*left*), and hypothetical conformations of hydrogen bonding between the copolymers and water which reveal two kinds (*right*): the intramolecular hydrogen bond among the carboxylic acid groups in the polymer chains when the pH is below p*K*a, and the intermolecular hydrogen bonds between PAA chains and water molecules when the pH is above p*K*a. Reproduced from Ref. [1] by permission of John Wiley & Sons Ltd

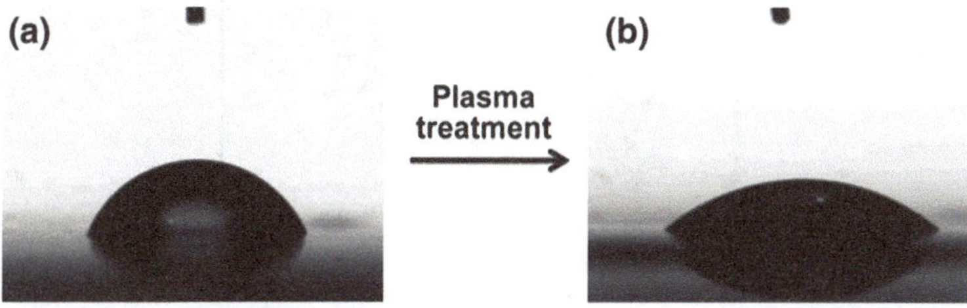

Fig. 3.5 The CA of the PET film surfaces before and after plasma treatment. **a** Before plasma treatment. **b** After plasma treatment. Reproduced from Ref. [1] by permission of John Wiley & Sons Ltd

PAA is a weak polyelectrolyte that adopts a coiled conformation depending on the degree of dissociation or protonation, which is controlled by the pH and ionic strength of the surrounding solution [26], due to ionic repulsion between anionic groups. PAA contains carboxylic groups that become ionized at pH values above its p*K*a of 4.7 (Fig. 3.4) [23]. The conformation of this weak polyelectrolyte changes from coiled at low pH to stretched at high pH [27]. The effect of pH on chain conformation on the surface is schematically outlined in Fig. 3.4 (right insert). At low pH values the formation of intramolecular hydrogen bonds in PAA prevails, which leads to a hydrophobic surface. However, in the case of a high pH value, intermolecular hydrogen bonds between water molecules and PAA were formed, and a hydrophilic surface was obtained.

Ionic transport properties of the nanochannel (Sample1) before and after plasma treatment have been examined by current measurements. Figure 3.6a shows *I–V* properties of the nanochannel before treatment, and the nanochannel exhibits linear *I–V* curves at different pH values. No change could be observed both when the pH changed from 5.8 to 2.8 and from 2.8 to 10, which meant the original nanochannel (Sample2) does not rectify at different pH values and without gating property (Fig. 3.7). After plasma treatment, there was a remarkable difference that the significant rectifications were observed as *I–V* curves (Fig. 3.6b). By changing the pH from 5.8 to 2.8, a significant decrease in the transmembrane ionic current was observed under the same ion concentration of the test solution (KCl 0.1 M). Then, a significant increase in the transmembrane ionic current could be observed when the pH changed from 2.8 to 10. It was thought that asymmetric ionic

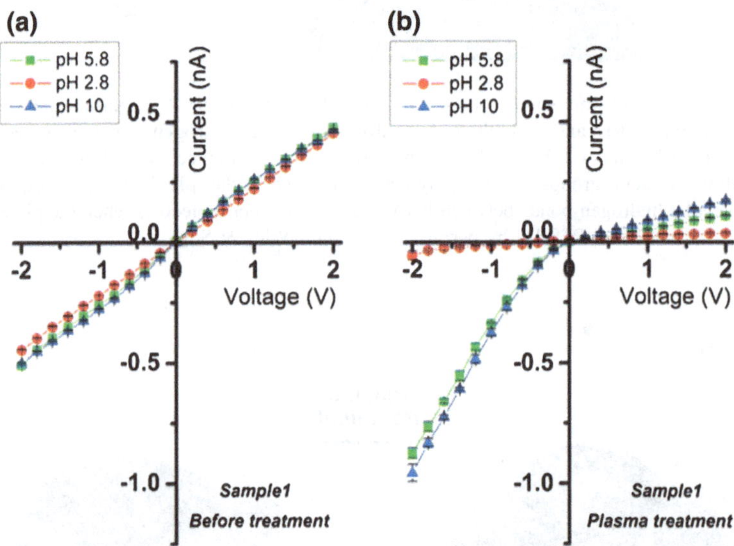

Fig. 3.6 a, b *I–V* properties of the single nanochannel under different pH conditions (pH 5.8, *green*; pH 2.8, *red*; pH 10, *blue*). (Sample1, base ~250 nm, tip ~11 nm, before plasma treatment) Reproduced from Ref. [1] by permission of John Wiley & Sons Ltd

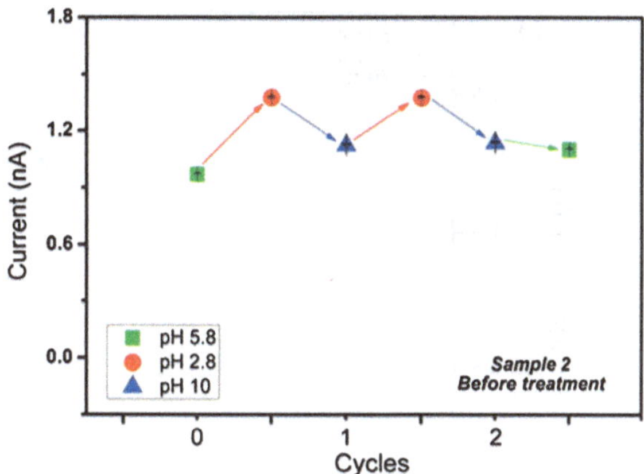

Fig. 3.7 Reversible variation of the ionic current transport of the single nanochannel before plasma treatment at 2 V (anode facing the plasma treatment side of the nanochannel, pH 5.8, *green*; pH 2.8, *red*; pH 10, *blue*). (Sample2, base ~270 nm, tip ~30 nm, before plasma treatment). Reproduced from Ref. [1] by permission of John Wiley & Sons Ltd

transport property was caused by asymmetric chemical modification, which brings the structural asymmetry and the non-uniform chemical composition of the nanochannel.

The open/close switching ability of the smart nanochannel system upon alternating the pH of the test solution between 2.8 and 10 (start and end at pH 5.8) is shown in Fig. 3.8a (Sample2), which reflects reproducible and reversible abilities of the pH-gating ionic transport nanodevice. The ionic transport property of this system changed immediately after the change in pH value of the test solution. The degree to which the ionic transport can be controlled depends on the size of the original nanochannel and the degree of plasma treatment. Some previous studies [5–8, 10, 26, 28–30] related to responsive nanochannels have shown that there were three major influencing factors bringing the change of ionic current at the same ion concentration and a certain voltage: surface charge [5–8, 30], the effective channel size [5, 10, 28, 29], and the wettability [8, 25] of the nanochannel.

In our strategies, the asymmetric ionic transport property was caused by the non-homogeneity of the distribution of surface charge and the structural asymmetry of the channel after plasma treatment (Fig. 3.3). It is generally believed that the conformation of the PAA chains changes from coiled to stretch, which would plug the channels and result in the decrease in the effective channel size, and therefore the transmembrane ionic current is correspondingly reduced [29]. However, in our case, the transmembrane ionic current change is totally opposite, which indicates variations in the effective channel size are not the major influence during the pH change. These results coincided with the hypothesis of Azzaroni et al. [8], who concluded that an increase in hydrophobicity of the nanochannel could lead to mediation of ionic transport in the boundaries of the nanochannel wall, due to hindering of the formation of the mobile electrolyte layer. It is thought that the wettability of the nanochannel is the key factor in our system. Considering the dependence of water permeation on pH, the tip of the nanochannel carrying carboxylic groups permeated water at higher rates under high pH conditions, in terms of hydrophilic state. This behavior can be simply explained by the

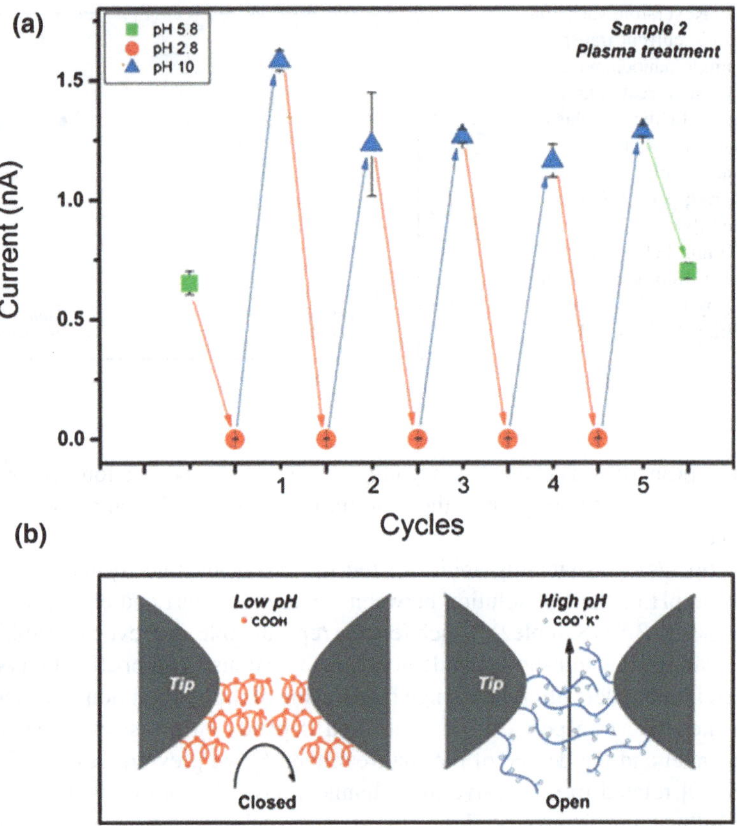

Fig. 3.8 **a** Reversible variation of the ionic current transport of the single nanochannel system at 2 V (anode facing the plasma treatment side of the nanochannel, pH 5.8, *green*; pH 2.8, *red*; pH 10, *blue*). (Sample2). **b** Explanation of pH-dependant water permeation through the single hour-glass shaped nanochannel. Reproduced from Ref. [1] by permission of John Wiley & Sons Ltd

mechanism shown in Fig. 3.8b. To be clearer, when the pH changes from low to high, the wettability of the nanochannel plays a prominent role in the ionic transport properties of the smart nanochannel because of the PAA transition from a hydrophobic state to a hydrophilic state.

3.5 Conclusions

In summary, we experimentally demonstrate a smart nanochannel material, which displays the advanced feature of providing simultaneous control over the pH-tunable asymmetric and pH-gating ion transport properties. Moreover, our simple plasma approach could provide an alternative to asymmetric chemical modification with various complicated functional molecules on the wall of the single nanochannels, in order to simulate different ionic transport processes as well as to enhance the functionality of ion channels.

References

1. Hou X, Liu Y, Dong H, Yang F, Li L, Jiang L (2010) A pH-gating ionic transport nanodevice: asymmetric chemical modification of single nanochannels. Adv Mater 22(22):2440–2443. doi:10.1002/adma.200904268
2. Hille B (2001) Ion channels of excitable membranes. Sinauer Associates, Sunderland
3. Rothman JE, Lenard J (1977) Membrane asymmetry. Science 195(4280):743–753
4. Hou X, Jiang L (2009) Learning from nature: building bio-inspired smart nanochannels. ACS Nano 3(11):3339–3342. doi:10.1021/Nn901402b
5. Xia F, Guo W, Mao YD, Hou X, Xue JM, Xia HW, Wang L, Song YL, Ji H, Qi OY, Wang YG, Jiang L (2008) Gating of single synthetic nanopores by proton-driven DNA molecular motors. J Am Chem Soc 130(26):8345–8350. doi:10.1021/Ja800266p
6. Ali M, Ramirez P, Mafe S, Neumann R, Ensinger W (2009) A pH-tunable nanofluidic diode with a broad range of rectifying properties. ACS Nano 3(3):603–608. doi:10.1021/nn900039f
7. Yameen B, Ali M, Neumann R, Ensinger W, Knoll W, Azzaroni O (2009) Single conical nanopores displaying pH-tunable rectifying characteristics. manipulating ionic transport with zwitterionic polymer brushes. J Am Chem Soc 131(6):2070–2071. doi:10.1021/ja8086104
8. Yameen B, Ali M, Neumann R, Ensinger W, Knoll W, Azzaroni O (2009) Synthetic proton-gated ion channels via single solid-state nanochannels modified with responsive polymer brushes. Nano Lett 9(7):2788–2793. doi:10.1021/nl901403u
9. Kalman EB, Vlassiouk I, Siwy ZS (2008) Nanofluidic bipolar transistors. Adv Mater 20(2):293–297. doi:10.1002/adma.200701867
10. Hou X, Guo W, Xia F, Nie FQ, Dong H, Tian Y, Wen LP, Wang L, Cao LX, Yang Y, Xue JM, Song YL, Wang YG, Liu DS, Jiang L (2009) A biomimetic potassium responsive nanochannel: G-Quadruplex DNA conformational switching in a synthetic nanopore. J Am Chem Soc 131(22):7800–7805. doi:10.1021/Ja901574c
11. Wang G, Bohaty AK, Zharov I, White HS (2006) Photon gated transport at the glass nanopore electrode. J Am Chem Soc 128(41):13553–13558. doi:10.1021/ja064274j
12. Yameen B, Ali M, Neumann R, Ensinger W, Knoll W, Azzaroni O (2009) Ionic transport through single solid-state nanopores controlled with thermally nanoactuated macromolecular gates. Small 5(11):1287–1291. doi:10.1002/smll.200801318
13. Nishizawa M, Menon VP, Martin CR (1995) Metal nanotubule membranes with electrochemically switchable ion-transport selectivity. Science 268(5211):700–702. doi:10.1126/science.268.5211.700
14. Ali M, Yameen B, Neumann R, Ensinger W, Knoll W, Azzaroni O (2008) Biosensing and supramolecular bioconjugation in single conical polymer nanochannels. Facile incorporation of biorecognition elements into nanoconfined geometries. J Am Chem Soc 130(48):16351–16357. doi:10.1021/ja8071258
15. Li NC, Yu SF, Harrell CC, Martin CR (2004) Conical nanopore membranes preparation and transport properties. Anal Chem 76(7):2025–2030. doi:10.1021/ac035402e
16. Buyukserin F, Kang M, Martin CR (2007) Plasma-etched nanopore polymer films and their use as templates to prepare “nano test tubes”. Small 3(1):106–110. doi:10.1002/smll.200600267
17. Ito Y, Park YS (2000) Signal-responsive gating of porous membranes by polymer brushes. Polym Adv Technol 11(3):136–144. doi:10.1002/1099-1581(200003)11:3<136:aid-pat961>3.0.co;2-f

18. Dmitriev SN, Kravets LI, Sleptsov VV, Elinson VM (2005) Plasma-induced graft polymerisation of 2-methyl-5-vinylpyridine on the surface of poly(ethylene terephthalate) track membranes. Polym Degrad Stab 90(2):374–378. doi:10.1016/j.polymdegradstab.2004.11.026
19. Xie R, Chu LY, Chen WM, Xiao W, Wang HD, Qu JB (2005) Characterization of microstructure of poly (*N*-isopropylacrylamide)-grafted polycarbonate track-etched membranes prepared by plasma-graft pore-filling polymerization. J Membr Sci 258(1–2):157–166. doi:10.1016/j.memsci.2005.03.012
20. Apel P (2001) Track etching technique in membrane technology. Radiat Meas 34(1–6):559–566. doi:10.1016/s1350-4487(01)00228-1
21. Xie Y, Wang X, Xue J, Jin K, Chen L, Wang Y (2008) Electric energy generation in single track-etched nanopores. Appl Phys Lett 93(16). doi:163116 10.1063/1.3001590
22. Wharton JE, Jin P, Sexton LT, Horne LP, Sherrill SA, Mino WK, Martin CR (2007) A method for reproducibly preparing synthetic nanopores for resistive-pulse biosensors. Small 3(8):1424–1430. doi:10.1002/smll.200700106
23. Lee JW, Kim SY, Kim SS, Lee YM, Lee KH, Kim SJ (1999) Synthesis and characteristics of interpenetrating polymer network hydrogel composed of chitosan and poly(acrylic acid). J Appl Polym Sci 73(1):113–120. doi:10.1002/(sici)1097-4628(19990705)73:1<113::aid-app13>3.3.co;2-4
24. Xia F, Feng L, Wang ST, Sun TL, Song WL, Jiang WH, Jiang L (2006) Dual-responsive surfaces that switch superhydrophilicity and superhydrophobicity. Adv Mater 18(4):432–436. doi:10.1002/adma.200501772
25. Yang SC (2006) Effects of surface roughness and interface wettability on nanoscale flow in a nanochannel. Microfluid Nanofluid 2(6):501–511. doi:10.1007/s10404-006-0096-5
26. Carroll T, Booker NA, Meier-Haack J (2002) Polyelectrolyte-grafted microfiltration membranes to control fouling by natural organic matter in drinking water. J Membr Sci 203(1–2):3–13. doi:10.1016/s0376-7388(01)00701-3
27. Bajpai AK (1997) Interface behaviour of ionic polymers. Prog Polym Sci 22(3):523–564. doi:10.1016/s0079-6700(96)00003-2
28. Harrell CC, Kohli P, Siwy Z, Martin CR (2004) DNA—Nanotube artificial ion channels. J Am Chem Soc 126(48):15646–15647. doi:10.1021/ja044948v
29. Wang Y, Liu ZM, Han BX, Dong ZX, Wang JQ, Sun DH, Huang Y, Chen GW (2004) pH Sensitive polypropylene porous membrane prepared by grafting acrylic acid in supercritical carbon dioxide. Polymer 45(3):855–860. doi:10.1016/j.polymer.2003.11.042
30. Wanunu M, Meller A (2007) Chemically modified solid-state nanopores. Nano Lett 7(6):1580–1585. doi:10.1021/nl070462b

Chapter 4
Asymmetric Temperature/pH Dual-Responsive Symmetric Hour-Glass Shaped Single Nanochannel

4.1 Introduction

Artificial single nanochannels have emerged as possible candidates for mimicking the process of ionic transport in ion channel and boosting the development of bio-inspired intelligent nanomachines for real-world applications, such as biosensors, molecular filtration, and nanofluidic devices. One challenge that remains is to make the artificial nanochannel "smart" with various functions like organism in nature. The components of ion channels are asymmetrically distributed between membrane surfaces, which are significant for the implementation of the complex biological function. Inspired by this natural asymmetrical design, here I introduce a biomimetic asymmetric responsive single nanochannel system that displays the advanced feature of providing control over pH and temperature cooperation tunable asymmetric ionic transport property through asymmetric modifications inside the single nanochannels, which could be considered as a primary platform for the simulation of the different ionic transport processes as well as the enhancement of the functionality of ion channels [1].

Living cells depend on ion channels to communicate chemically and electrically with the extracellular environment [2]. The ability to simulate the ionic transport process of ion channels permits a better understanding of how ion channels work and provides tools for many applications in life science and materials science [3–5]. One approach is to use an artificial nanochannel [6, 7] of known with greater flexibility in terms of shape and size, superior robustness, and surface properties, which can be tuned depending on the desired function, to mimic the biological channels [8, 9]. There has been rapid progress in developing the artificial nanochannels that respond to a single external stimulus, such as specific ions [10–12], light [13, 14], pH [15–21], temperature [22–24], and mechanical stress [25]. However, all of these nanochannels are responsive to only one kind of external stimuli, and how to endow these artificial nanochannels with more intelligence is still a challenging task [7, 26].

In this work, I further extend the function of biomimetic nanochannel systems by using an asymmetric chemical modification approach [20]. Compared with

X. Hou, *Bio-inspired Asymmetric Design and Building of Biomimetic Smart Single Nanochannels*, Springer Theses,
DOI: 10.1007/978-3-642-38050-1_4, © Springer-Verlag Berlin Heidelberg 2013

previous systems [15–24], this asymmetric responsive nanochannel system has the advantage that it provides simultaneous control over both the pH- and temperature-tunable asymmetric ionic transport properties. Such a system, as an example, could potentially spark further experimental and theoretical efforts with different complicated functional molecules to exploit more complex "smart" nanochannel systems.

The shape [27] and chemical properties [28] of the nanochannels are two key factors to control ionic transport properties inside the nanochannels. Once artificial nanochannels are prepared, it is difficult to change them within a wide range of shape. Therefore, chemical modification of the interior surface of the nanochannels with functional molecules is a critical component in the advancement of smart functional nanochannels. It is because functional modification of the channel can change both sizes and chemical properties of the nanochannels at their narrowest point.

In order to achieve different functionalities of the interior surface of artificial nanochannels, various methods have been invented, such as electroless deposition [29], solution chemical modification [16, 30], electrostatic self-assembly [31], and coverage of the whole inner surface of the nanochannels. Recently, Jiang and others have successfully developed temperature-controllable nanochannels with thermally nanoactuated macromolecular gates using electroless deposition and solution chemical modification approaches [22–24].

As we know, the components of most biological nanochannels are asymmetrically distributed between membrane surfaces to implement complex biological functions [32, 33]. Inspired by these asymmetrical nanochannels, different asymmetric chemical modification approaches have been developed to enhance the functionality of artificial nanochannels, such as electron beam evaporation [34], ion sputtering deposition [35], and plasma modification [20, 36]. In the previous chapter, I have developed a pH-controllable nanochannel which displays the advanced feature of providing simultaneous control over the pH-tunable asymmetric and pH-gated ionic transport properties by using a plasma asymmetric chemical modification approach [20]. On the basis of the prior work, here I further utilize the specific symmetric shape of a single nanochannel with asymmetric chemical modification approaches to functionalize diverse specific local areas with different functional molecules at nanometer dimensions in order to develop an asymmetric dual-responsive nanochannel system that provides simultaneous control over both the temperature and pH-tunable asymmetric ionic transport properties. Compared with the single external stimulus systems, the present work is more complicated and moves one step further toward the development of "smart" nanochannel systems for real-world applications.

As shown in Fig. 4.1, we prepared a single nanochannel membrane with the well-developed ion-track-etching technology [37, 38]. The nanochannel was embedded in a track-etched PET membrane with a single ion track in the center. The track-etching technique allows control over the shape of the channels, and in this work the etched single nanochannel is symmetric and hour-glass shaped. Diameter measurements of hour-glass shaped nanochannels were conducted with a

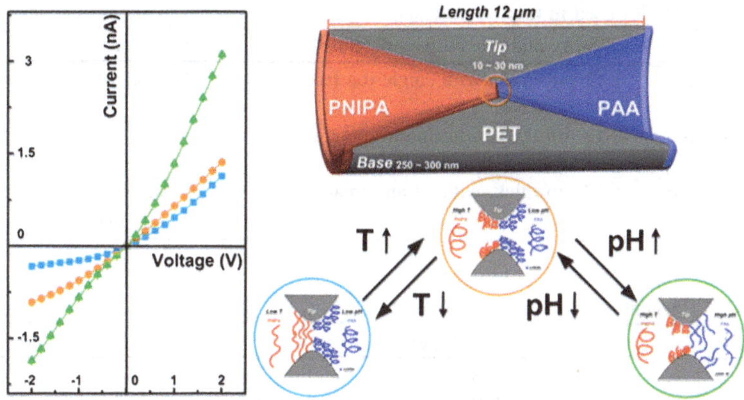

Fig. 4.1 Scheme of a biomimetic asymmetric temperature/pH-responsive single nanochannel. Reprinted with the permission from Ref. [1]. Copyright 2010 American Chemical Society

commonly used electrochemical method [39]. The opening at the base was usually ~250–300 nm wide, and the narrow center (tip) was ~10–30 nm wide. One side of the nanochannel was treated by plasma-induced grafting in the vapor phase of *N*-isopropylacrylamide (NIPA), which became a temperature-responsive polymer [40], poly (*N*-isopropylacrylamide) (PNIPA), after plasma-induced graft polymerization. The other side of the nanochannel was treated by plasma-induced grafting in the vapor phase of distilled AAc, which became a pH-responsive polymer [20], PAA, after plasma-induced graft polymerization. After grafting PNIPA on one side of the hour-glass shaped nanochannel wall and PAA on the other side, the PNIPA underwent thermally responsive formation of intermolecular hydrogen-bonding between PNIPA chains and water molecules and intramolecular hydrogen-bonding between C=O and N–H groups in PNIPA chains below and above the lower critical solution temperature (LCST) of about 32 °C [40]. PAA underwent pH-responsive formation of the intramolecular hydrogen bonds among the carboxylic acid groups in the polymer chains when the pH was below the p*K*a of 4.7 and of the intermolecular hydrogen bonds between PAA chains and water molecules when the pH was above the p*K*a [20, 41]. These asymmetric temperature/pH-responsive conformational conversions could well induce changes of ionic transport properties inside the nanochannel and thus realized the asymmetric dual-responsive smart nanochannel system.

4.2 Materials and Methods

4.2.1 Materials

See Tables 4.1 and 4.2.

Table 4.1 Samples used in the experiments

Sample	Purity and related parameters	Provider
PET membrane	12 μm thick, with single ion track in the center	GSI, Germany
PET membrane	12 μm thick, with multi-ions track in the center ($10^7/cm^2$)	
PET membrane	12 μm thick, without any treatment	Hostaphan RN12 Hoechst, Germany

Table 4.2 Reagents used in the experiments

Reagent	Purity quotient	Provider
KCl	A.R.	Beijing Chemical Works
HCl	A.R.	Beijing Chemical Works (contents 36–38 %)
NaOH	A.R.	Beijing Chemical Works
KOH	A.R.	West Long Chemical Co., Ltd., China
HCOOH	A.R.	Beijing Chemical Works (contents >88 %)
Acrylic acid (AAc)	A.R.	Aladdin, China
N-isopropylacrylamide (NIPA)	E.P.	Tokyo Chemical Industry Co., Ltd., Japan
Liquid paraffin	C.P.	West Long Chemical Co., Ltd., China
Filamentary silver	99.99 %	Made in China
Platinum filament	99.99 %	Made in China

4.2.2 Instruments

See Table 4.3.

4.2.3 Solvent Preparation

Track-etching solution for the nanochannel preparation: 9 M NaOH
Stopping solution for the etching solution: 1 M KCl + 1 M HCOOH
Transmembrane current test solution: KCl (0.1 M and 1 M)

4.3 Experiment Operation

4.3.1 Plasma-Induced Graft Polymerization

The PET film should first be soaked in water for 5 h after the etching experiment. There are two steps for our asymmetric chemical modification approach. First step, one side of the nanochannel was modified with PNIPA. NIPA was placed in a

Table 4.3 Instruments used in the experiments

Instrument	Producer
Water purification system (water, 18.2 MΩ)	Milli-Q, Millipore Corporation, USA
KEITHLEY 6487 picoammeter	Keithley Instruments, Cleveland, OH, USA
DT-01 and DT-02 Plasma instrument	SuZhou OPS Plasma Technology Co., Ltd., China
DJ-02, Plasma graft instrument and homemade temperature control device (including liquid paraffin)	SuZhou OPS Plasma Technology Co., Ltd., China and ICCAS, China
ZF-7 Uviol lamp	Shanghai Gucun Optic Instrument Factory, China
Model PHS-25 pH meter instruction	Shanghai Precision & Scientific Instrument Co., Ltd.
Field emission scanning electron microscope, JSM-6700F	JEOL, Japan
X-ray photoelectron spectroscopy, ESCALab220i-XL	Thermo VG Scientific Ltd., UK
Homemade PTFE electrolyzer	ICCAS, China
Homemade plexiglass electrolyzer	ICCAS, China
Hot plate clarkson H3400-HS07	IKA, USA
Stopwatch timer	Made in China
Electronic thermometer	Made in China

monomer delivery system (DJ-02). Vacuum before switching on the glow discharge was 5 Pa, and the working temperature was maintained at 92 °C, due to the low volatility of NIPA. In my experiment setup, the loop of pumping and releasing argon gas in the reaction chamber (DT-02) runs three times. NIPA monomers were then transported to the reaction chamber, and it was maintained at a power of 20 W to glow discharge in the reaction chamber. At the same time, the argon atmosphere was kept about 40–60 Pa, and this process lasted for 15 min. After the glow extinguished, it continued for about 20 min under vacuum of 40–60 Pa for the grafting reaction. After that, the chamber was connected with air. For the second step, the other side of the nanochannel was modified with PAA. Distilled AAc was injected into plasma-induced grafting reactor (DJ-01). Vacuum before switching on the glow discharge was 21 Pa, and the working temperature was 20 °C. In my experiment setup, the argon atmosphere was kept about 50–70 Pa, and the process lasted for 15 min. It continued about 10 min for Start R-F power supply source (DT-01) at 20 W to glow discharge. After the glow extinguished, the grafting reactor would lead to grafting AAc monomers, maintaining the vacuum at 150–300 Pa. The grafting reaction lasted about 20 min. After that, the chamber was connected with air. The plasma treatment was finished. The following experimental results of the asymmetric responsive ionic transport properties of the channel system are the indirect evidence of the successful asymmetric chemical modification of the channel.

4.3.2 Current Measurement

The ionic transport properties of the nanochannel were studied by measuring ionic current through the nanochannels before and after plasma treatment. Ionic current was measured by a Keithley 6487 picoammeter. An hour-glass shaped single nanochannel PET membrane was mounted between two chambers of the etching cell mentioned above. Ag/AgCl electrodes were used to apply a transmembrane potential across the film. Forward voltage was the potential applied on one of base sides, which is the PNIPA side of the nanochannel. The main transmembrane potential used in this work was evaluated, and a scanning voltage varied from −2 to +2 V with a 40 s period was selected. The pH of the electrolyte was adjusted by 1 M HCl and KOH solutions, and the influence of addition substance quality can be ignored. It should be clear that all the pH values in this work are mensurated at 20 °C. All measurements were carried out in a custom-designed temperature control system. In this work, each test was repeated five times to obtain the average current value at different voltages on the same channel.

4.3.3 XPS Testing

XPS measurements were performed to validate the chemical identity of the PET films before and after PNIPA chemical modification. XPS data were obtained by an ESCALab220i-XL electron spectrometer from VG Scientific using 300 W Al Kα radiation, and the base pressure was about 3×10^{-9} mbar. The binding energies were referenced to the C1s line at 284.8 eV from adventitious carbon. The original PET samples were treated with 9 M NaOH for 15 min. The samples were then taken out from the etching solution and treated with the stopping solution (1 M KCl + 1 M HCOOH) for half an hour. After that, the samples were treated with the deionized water, and had been stored for 5 h in the deionized water before further experiments. The PET film was treated by plasma-induced grafting in the vapor phase of the *N*-isopropylacrylamide, which became a temperature-responsive polymer (PNIPA), after plasma-induced graft polymerization. After modification, the sample was treated with the deionized water, and had been stored for one day. Before the XPS test, the sample was blown dry by N_2. Figure 4.2 shows the XPS spectra of PET films before and after PNIPA modification. In Fig. 4.2b, it is clear that there is an obvious peak near 400, which indicating a typical “N1s”, thus it simply indicates PNIPA was immobilized on the surface of the PET film.

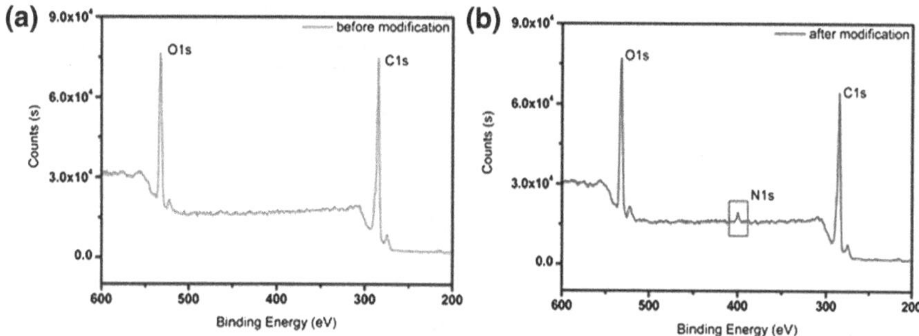

Fig. 4.2 XPS spectra of PET films before and after PNIPA modification. **a** Before modification. **b** After modification. Reprinted with the permission from Ref. [1]. Copyright 2010 American Chemical Society

4.4 Results and Discussion

Ionic transport properties of the single nanochannel before and after PNIPA/PAA asymmetric chemical modification have been examined by current measurements. Figure 4.3a shows *I–V* properties of the nanochannel before chemical modification, and the original nanochannel exhibits linear *I–V* curves with temperature from 23 °C to 40 °C at pH 2.8. The degree of ionic rectification was defined as the $ratio_{+/-}$ of absolute values of ionic currents recorded at a given positive voltage (+2 V, anode facing the PNIPA side of the nanochannel) and at the same absolute value of a negative voltage (−2 V, anode facing the PAA side of the nanochannel). Before modification, the $ratio_{+/-}$ ($\sim$1.0) stayed nearly unchanged with temperature from 23 to 40 °C.

After modification, there was an obvious difference, in that the rectifications were observed as asymmetric *I–V* curves (Fig. 4.3b). By changing the temperature from 23 to 40 °C, the $ratio_{+/-}$ changed from 3.61 ± 0.17 to 1.52 ± 0.03, and the trend of ratio decreasing coincided with the previous study on the temperature-responsive conical shaped metalpolymer composite single nanochannel [24]. It is worth mentioning that there was a remarkable asymmetric ionic current increase when the temperature was changed from 23 to 32 °C, compared with changing the temperature from 32 to 40 °C. This behavior can be explained by the mechanism shown in Fig. 4.3c.

It is generally believed that PNIPA maintains a swollen state at 23 °C and exhibits a thermally responsive transition into a collapsed state below and above LCST of about 32 °C [40]. Raising the temperature from 23 to 32 °C promotes drastic changes in the conformational state of the PNIPA. In this case, there is an obvious increase of the effective channel size of the nanochannel, which is evidenced as a larger increase in ionic current. On continuing to raise the temperature from 32 to 40 °C, the conformational state of the PNIPA stayed nearly unchanged. In this case, there is no increase of the effective channel size of the nanochannel, and the ionic current increase coming from the conductivity changes due to

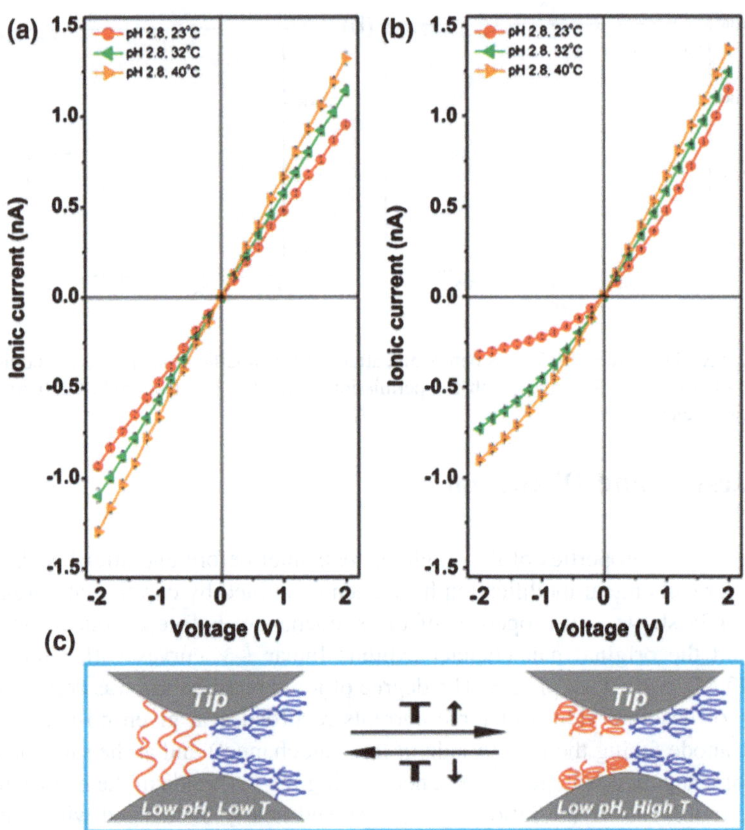

Fig. 4.3 *I–V* properties of the single nanochannel **a** before and **b** after PNIPA attached onto one side of the inner channel wall, and PAA attached onto the other side (asymmetric chemical modification). *I–V* characteristics were recorded under symmetric electrolyte conditions (pH 2.8) at 23 °C (*red circles*), 32 °C (*green triangles*), and 40 °C (*orange triangles*). **c** Explanation of the asymmetric dual-responsive ionic transport properties of this hour-glass shaped single nanochannel system at low pH with different temperatures (sample, base ~290 nm, tip ~18 nm, before chemical modification). Reprinted with the permission from Ref. [1]. Copyright 2010 American Chemical Society

temperature variations, the same as ionic transport properties of the original nanochannel. It is clear that asymmetric ionic transport property was caused by the asymmetric PNIPA modification, which gives structural asymmetry and nonuniform chemical composition to the nanochannel. This nanochannel system has the advantage that it provides simultaneous thermally controllable and asymmetric ionic transport properties that the other temperature-responsive nanochannels [23] do not have.

Figure 4.4a shows *I–V* properties of the nanochannel before chemical modification, and the original nanochannel exhibits linear *I–V* curves with pH from 2.8 to 10 °C at 23 °C. The $\text{ratio}_{+/-}$ (~1.0) stayed nearly unchanged with pH from 2.8 to

10, which meant that the original nanochannel does not rectify at different pH values. After modification, there was a remarkable difference: a significant increase in the transmembrane ionic current could be observed when the pH changed from 2.8 to 10 in Fig. 4.4b. By changing the pH from 2.8 to 10, the ratio +/− changed from 3.6 ± 0.17 to 1.86 ± 0.01. According to our previous pH-controllable nanochannel work [20], it was thought that this pH-responsive ionic transport property was caused by asymmetric PAA modification, and the pH-responsive capabilities can be controlled, depending on the size of the original nanochannel and the degree of plasma modification. It was assumed that during the pH changes from low to high, the wettability of the channel plays a prominent role in the ionic transport properties because of the PAA transition from a hydrophobic state to a hydrophilic state. This behavior can be explained by the mechanism shown in Fig. 4.4c.

It was assumed that the effective channel size of the nanochannel was determined by two aspects in this system, that is, the changes of physical block and charge density caused by the molecular conformational change [11]. In our system, the physical block and charge density would change simultaneously because the PAA molecules were negatively charged when pH was above p*K*a. Thus, these two factors were integrated to investigate the change of "effective channel size", which could more comprehensively explain the phenomenon in the asymmetric responsive system.

Raising the temperature, compared with PAA, the conformation of the PNIPA chains changes from the swollen state to the collapsed state, and there is also a conformational transition that changes the wettability of the nanochannel. The PNIPA transition, in contrast, is from a hydrophilic state to a hydrophobic state. The phenomenon of the responsive properties of the nanochannel is explained by the competition between two factors, that is, the changes of the effective channel size and the wettability. It is reasonable that when PNIPA transformed from the swollen state to the collapsed state, the effective channel size and the wettability of the nanochannel changed at the same time. PNIPA neutralized the surface charge of the nanochannel, resulting in asymmetric surface charge on the inner surface of the symmetric nanochannel, and the increase of the effective channel size of nanochannel played a prominent role in ionic transport properties. It is worth mentioning that the pH-responsive PAA molecules have negatively charged carboxyl groups. When the pH is above the p*K*a, PAA molecules were highly negatively charged. This property might effectively slow down the decrease of the effective channel size of the nanochannel during the change in the conformation of the PAA chains from coiled to stretch, due to mutual electrostatic repulsion between PAA charged molecules. In this case, it is clear that PNIPA and PAA would change simultaneously, and the pH responsivity is, of course, greater than the temperature responsivity.

Figure 4.5a shows *I–V* properties of the nanochannel before chemical modification, and the $\text{ratio}_{+/-}$ (~ 1.1) stayed nearly unchanged with temperature from 23 to 40 °C. After modification, the *I–V* curves were asymmetric (Fig. 4.5b), and the ratio +/− changed from 1.86 ± 0.01 to 1.70 ± 0.02 upon changing the

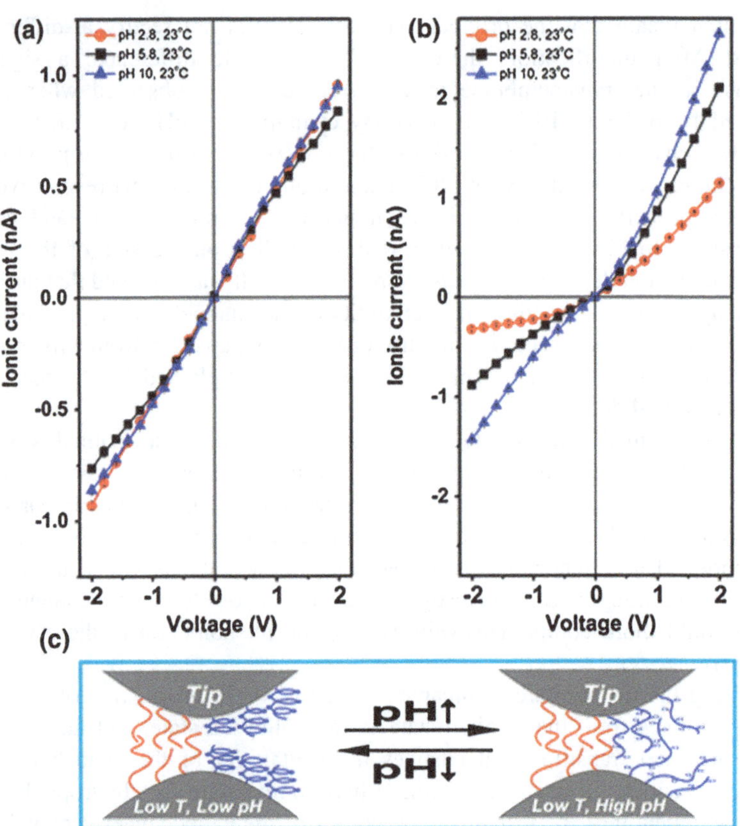

Fig. 4.4 *I–V* properties of the single nanochannel **a** before and **b** after asymmetric chemical modification. *I–V* characteristics were recorded under symmetric electrolyte conditions (23 °C) at pH 2.8 (*red circles*), pH 5.8 (*black squares*), and pH 10 (*blue triangles*). **c** Explanation of the asymmetric dual-responsive ionic transport properties of this hour-glass shaped single nanochannel system at low temperature with different pH. Reprinted with the permission from Ref. [1]. Copyright 2010 American Chemical Society

temperature from 23 to 40 °C. As mentioned above, the influence of pH is greater than that of temperature. Although the effective channel size increased with the increase of temperature at high pH, that phenomena was not observed at the low pH state: the ratio change is obvious, and the asymmetric ionic current increases with temperature increasing from 23 to 32 °C. This behavior can be explained by the mechanism shown in Fig. 4.5c.

Figure 4.6a shows *I–V* properties of the original nanochannel, which exhibits linear *I–V* curves with pH from 2.8 to 10 at 40 °C. The $\text{ratio}_{+/-}$ (~ 1.0) stayed nearly unchanged with pH from 2.8 to 10. Compared to the low-temperature state, the nanochannel after modification showed a similar increasing trend with pH from 2.8 to 10, which was also observed as asymmetric *I–V* curves (Fig. 4.6b). However, the $\text{ratio}_{+/-}$ significantly decreased with the increasing temperature from 23 to 40 °C. As

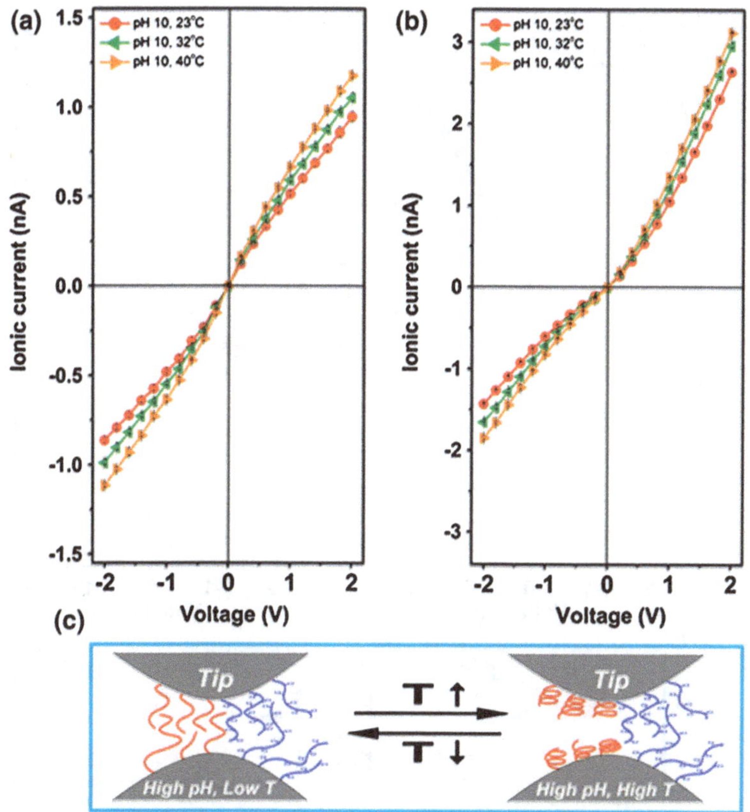

Fig. 4.5 *I–V* properties of the single nanochannel **a** before and **b** after asymmetric chemical modification. *I–V* characteristics were recorded under symmetric electrolyte conditions (pH 10) at 23 °C (*red circles*), 32 °C (*green triangles*), and 40 °C (*orange triangles*). **c** Explanation of the asymmetric dual-responsive ionic transport properties of this hour-glass shaped single nanochannel system at high pH with different temperatures. Reprinted with the permission from Ref. [1]. Copyright 2010 American Chemical Society

mentioned above, there is an obvious increase of the effective channel size of the nanochannel upon changing the temperature from 23 to 40 °C, and it is generally believed that the ionic rectification decreases with the increase of the nanometer dimensions of the effective channel size induced by ambient stimuli [42]. This behavior can be explained by the mechanism shown in Figs. 4.4c and 4.6c.

As shown in Fig. 4.7a, the ionic current in the nanochannel before modification stayed nearly unchanged at different temperatures and pH, which meant that the original nanochannel does not rectify. In our smart nanochannel system, there is a negative correlation between the ratio$_{+/-}$ and the temperature at various pH values (Fig. 4.7b). This property could be attributed to the thermally responsive PNIPA transition from the swollen state to the collapsed state, which induced the change of the effective channel size of the nanochannel. In addition, at low pH state, the

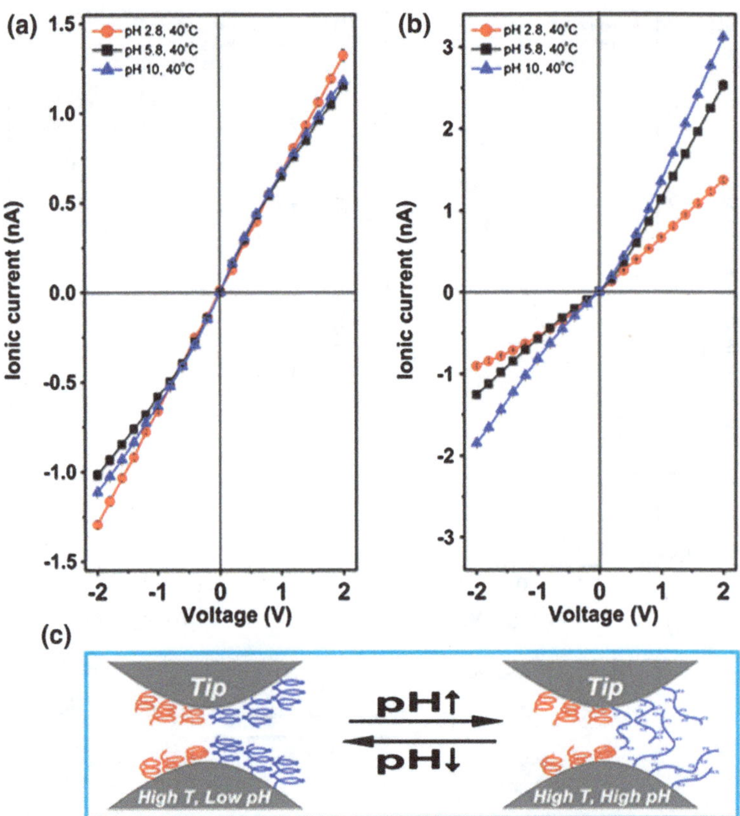

Fig. 4.6 *I–V* properties of the single nanochannel **a** before and **b** after asymmetric chemical modification. *I–V* characteristics were recorded under symmetric electrolyte conditions (40 °C) at pH 2.8 (*red circles*), pH 5.8 (*black squares*), and pH 10 (*blue triangles*). **c** Explanation of the asymmetric dual-responsive ionic transport properties of this hour-glass shaped single nanochannel system at high temperature with different pH. Reprinted with the permission from Ref. [1]. Copyright 2010 American Chemical Society

change of this negative correlation is remarkable. It can be concluded that when the pH is below the p*K*a, PAA molecules were uncharged, the contribution of PAA to ionic transport properties was largely reduced, and the contribution of PNIPA response played a prominent role in asymmetric ionic transport properties of the nanochannel by asymmetric adjustment of the effective channel size, whereas when the pH was above the p*K*a, PAA molecules were highly negatively charged and the contribution of PAA started to exert a main influence on the ionic transport properties of the nanochannel.

To explore the asymmetric temperature-responsive properties with various pH, the degree of temperature influence on ionic transport properties of the nanochannel was defined as the ratio_T of values of ionic currents recorded at a high-temperature state (40 °C) as T_{high} and a low-temperature state (23 °C) as T_{low}. As

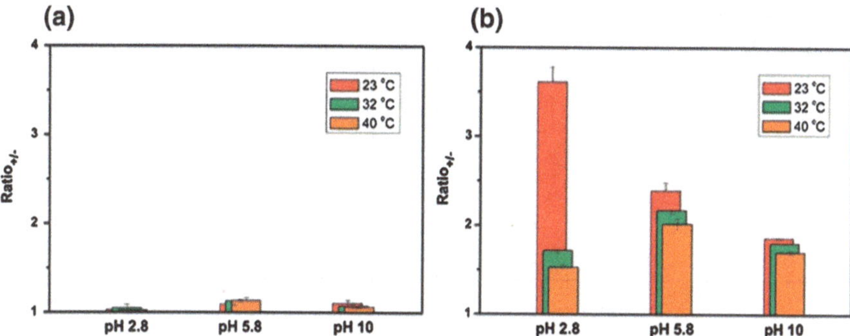

Fig. 4.7 Ionic current rectification of the single nanochannel **a** before and **b** after asymmetric chemical modification (23 °C, *red*; 32 °C, *green*; 40 °C, *orange*) at 2 V. Reprinted with the permission from Ref. [1]. Copyright 2010 American Chemical Society

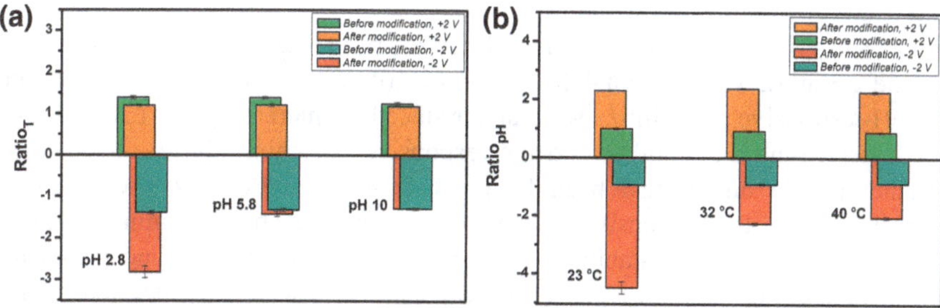

Fig. 4.8 Asymmetric responsive ionic transport properties of the nanochannel. **a** Current ratio of a high-temperature (40 °C) state versus a low-temperature (23 °C) state calculated in the single nanochannel before (*green* and *cyan*) and after (*orange* and *red*) asymmetric chemical modification with positive potential +2 V (anode facing the PNIPA side of the nanochannel, *above*) and negative potential −2 V (anode facing the PAA side of the nanochannel, *below*). **b** Current ratio of pH 10 state versus pH 2.8 state calculated in the single nanochannel before (*green* and *cyan*) and after (*orange* and *red*) asymmetric chemical modification with positive potential +2 V (*above*) and negative potential −2 V (*below*). Reprinted with the permission from Ref. [1]. Copyright 2010 American Chemical Society

shown in Fig. 4.8a, the ratio_T of values of the original nanochannel stayed around 1.3 at ±2 V with pH from 2.8 to 10. After modification, the ratio_T of values of the nanochannel was around 1.2 at +2 V, and from 2.83 ± 0.14 to 1.29 ± 0.01 at −2 V. This result again verified that the asymmetric responsive ionic transport property was caused by asymmetric chemical modification and the temperature-responsive capabilities could be reduced by increasing the pH.

Figure 4.8b shows the asymmetric pH-responsive properties with various temperatures. The degree of pH influence on ionic transport properties of the nanochannel was defined as the ratio_{pH} of values of ionic currents recorded at a high pH state (pH 10) as base and a low pH state (pH 2.8) as acid. The ratio_{pH} of values of the original nanochannel stayed around 0.9 at ±2 V with temperature

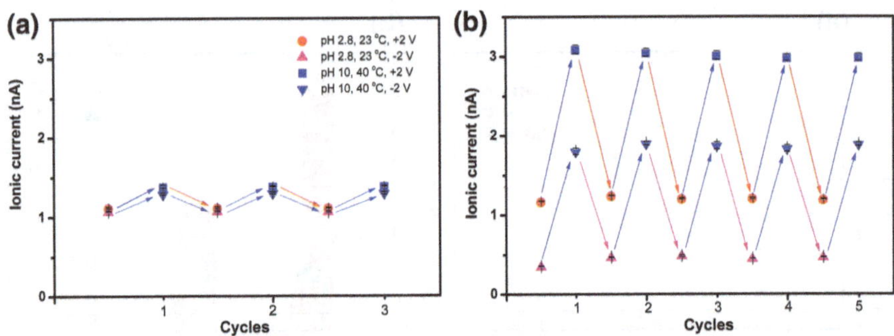

Fig. 4.9 Stability and responsive switching ability of the nanochannel. Reversible variation of the ionic current transport of the single nanochannel before **a** and after **b** asymmetric chemical modification at +2 V (pH 2.8 and 23 °C, *red circles*; pH 10 and 40 °C, *blue squares*) and −2 V (pH 2.8 and 23 °C, *magenta triangles*; pH 10 and 40 °C, *navy triangles*). Reprinted with the permission from Ref. [1]. Copyright 2010 American Chemical Society

from 23 to 40 °C. After modification, the ratio_T of values of the nanochannel was around 2.3 at +2 V, and from 4.48 ± 0.21 to 2.05 ± 0.04 at −2 V. It is clear that the pH-responsive capabilities were also reduced by increasing the temperature.

To determine the stability and the responsive switching ability of the nanochannel, further studies have been done and are shown in Fig. 4.9. The time scale of each cycle of the responsive switching experiment is about 30 min, due to the temperature control and stability. There is a stability of the original symmetric nanochannel (Fig. 4.9a), which is an ideal platform for developing smart nanochannel systems. The responsive switching ability of the smart nanochannel system upon alternating both the pH and the temperature of the test solution is shown in Fig. 4.9b, which reflects the reproducible and reversible character of the asymmetric responsive nanochannel.

4.5 Conclusions

In summary, a biomimetic asymmetric responsive single nanochannel system is experimentally demonstrated, which displays the advanced feature of providing simultaneous control over both temperature- and pH-tunable asymmetric ionic transport properties. The responsive properties of the smart nanochannel are explained by the competition between two factors: the changes of the effective channel size and the wettability. Increasing the temperature or the pH reduces the asymmetric responsive capabilities. We believe that this asymmetric system can be extended to higher levels of functionality through designing more complicated functional molecules. This novel system is as an example of the beginning of multiresponsive nanochannel systems and moves one step further toward the development of "smart" nanochannel systems for real-world applications [43–47].

References

1. Hou X, Yang F, Li L, Song Y, Jiang L, Zhu D (2010) A biomimetic asymmetric responsive single nanochannel. J Am Chem Soc 132(33):11736–11742. doi:10.1021/ja1045082
2. Hille B (2001) Ion channels of excitable membranes. Sinauer Associates, Sunderland
3. Martin CR, Siwy ZS (2007) Learning nature's way: biosensing with synthetic nanopores. Science 317(5836):331–332. doi:10.1126/science.1146126
4. Hou X, Jiang L (2009) Learning from nature: building bio-inspired smart nanochannels. ACS Nano 3(11):3339–3342. doi:10.1021/Nn901402b
5. Vlassiouk I, Apel PY, Dmitriev SN, Healy K, Siwy ZS (2009) Versatile ultrathin nanoporous silicon nitride membranes. Proc Natl Acad Sci USA 106(50):21039–21044. doi:10.1073/pnas.0911450106
6. Dekker C (2007) Solid-state nanopores. Nat Nanotechnol 2(4):209–215. doi:10.1038/nnano.2007.27
7. Baker LA, Bird SP (2008) Nanopores—a makeover for membranes. Nat Nanotechnol 3(2):73–74. doi:10.1038/nnano.2008.13
8. Alcaraz A, Ramirez P, Garcia-Gimenez E, Lopez ML, Andrio A, Aguilella VM (2006) A pH-tunable nanofluidic diode: electrochemical rectification in a reconstituted single ion channel. J Phys Chem B 110(42):21205–21209. doi:10.1021/jp063204w
9. Jung Y, Bayley H, Movileanu L (2006) Temperature-responsive protein pores. J Am Chem Soc 128(47):15332–15340. doi:10.1021/ja065827t
10. Powell MR, Sullivan M, Vlassiouk I, Constantin D, Sudre O, Martens CC, Eisenberg RS, Siwy ZS (2008) Nanoprecipitation-assisted ion current oscillations. Nat Nanotechnol 3(1):51–57. doi:10.1038/nnano.2007.420
11. Hou X, Guo W, Xia F, Nie FQ, Dong H, Tian Y, Wen LP, Wang L, Cao LX, Yang Y, Xue JM, Song YL, Wang YG, Liu DS, Jiang L (2009) A biomimetic potassium responsive nanochannel: G-Quadruplex DNA conformational switching in a synthetic nanopore. J Am Chem Soc 131(22):7800–7805. doi:10.1021/Ja901574c
12. Tian Y, Hou X, Wen L, Guo W, Song Y, Sun H, Wang Y, Jiang L, Zhu D (2010) A biomimetic zinc activated ion channel. Chem Commun 46(10):1682–1684. doi:10.1039/b918006k
13. Wang G, Bohaty AK, Zharov I, White HS (2006) Photon gated transport at the glass nanopore electrode. J Am Chem Soc 128(41):13553–13558. doi:10.1021/ja064274j
14. Wen LP, Hou X, Tian Y, Nie FQ, Song YL, Zhai J, Jiang L (2010) Bioinspired smart gating of nanochannels toward photoelectric-conversion systems. Adv Mater 22(9):1021–1024. doi:10.1002/adma.200903161
15. Schmuhl R, van den Berg A, Blank DHA, ten Elshof JE (2006) Surfactant-modulated switching of molecular transport in nanometer-sized pores of membrane gates. Angew Chem Int Edit 45(20):3341–3345. doi:10.1002/anie.200504579
16. Xia F, Guo W, Mao YD, Hou X, Xue JM, Xia HW, Wang L, Song YL, Ji H, Qi OY, Wang YG, Jiang L (2008) Gating of single synthetic nanopores by proton-driven DNA molecular motors. J Am Chem Soc 130(26):8345–8350. doi:10.1021/Ja800266p
17. Ali M, Ramirez P, Mafe S, Neumann R, Ensinger W (2009) A pH-tunable nanofluidic diode with a broad range of rectifying properties. ACS Nano 3(3):603–608. doi:10.1021/nn900039f
18. Yameen B, Ali M, Neumann R, Ensinger W, Knoll W, Azzaroni O (2009) Single conical nanopores displaying pH-tunable rectifying characteristics. Manipulating ionic transport with zwitterionic polymer brushes. J Am Chem Soc 131(6):2070–2071. doi:10.1021/ja8086104
19. Yameen B, Ali M, Neumann R, Ensinger W, Knoll W, Azzaroni O (2009) Synthetic proton-gated ion channels via single solid-state nanochannels modified with responsive polymer brushes. Nano Lett 9(7):2788–2793. doi:10.1021/nl901403u
20. Hou X, Liu Y, Dong H, Yang F, Li L, Jiang L (2010) A pH-gating ionic transport nanodevice: asymmetric chemical modification of single nanochannels. Adv Mater 22(22):2440–2443. doi:10.1002/adma.200904268

21. Yameen B, Ali M, Neumann R, Ensinger W, Knoll W, Azzaroni O (2010) Proton-regulated rectified ionic transport through solid-state conical nanopores modified with phosphate-bearing polymer brushes. Chem Commun 46(11):1908–1910. doi:10.1039/b920870d
22. Keyser UF, Koeleman BN, Van Dorp S, Krapf D, Smeets RMM, Lemay SG, Dekker NH, Dekker C (2006) Direct force measurements on DNA in a solid-state nanopore. Nat Phys 2 (7):473–477. doi: 10.1038/Nphys344
23. Yameen B, Ali M, Neumann R, Ensinger W, Knoll W, Azzaroni O (2009) Ionic transport through single solid-state nanopores controlled with thermally nanoactuated macromolecular gates. Small 5(11):1287–1291. doi:10.1002/smll.200801318
24. Guo W, Xia HW, Xia F, Hou X, Cao LX, Wang L, Xue JM, Zhang GZ, Song YL, Zhu DB, Wang YG, Jiang L (2010) Current rectification in temperature-responsive single nanopores. ChemPhysChem 11(4):859–864. doi:10.1002/cphc.200900989
25. Huh D, Mills KL, Zhu X, Burns MA, Thouless MD, Takayama S (2007) Tuneable elastomeric nanochannels for nanofluidic manipulation. Nat Mater 6(6):424–428. doi:10.1038/nmat1907
26. Savariar EN, Krishnamoorthy K, Thayumanavan S (2008) Molecular discrimination inside polymer nanotubules. Nat Nanotechnol 3(2):112–117. doi:10.1038/nnano.2008.6
27. Siwy ZS, Howorka S (2010) Engineered voltage-responsive nanopores. Chem Soc Rev 39(3):1115–1132. doi:10.1039/b909105j
28. Gyurcsanyi RE (2008) Chemically-modified nanopores for sensing. Trac-Trend Anal Chem 27(7):627–639. doi:10.1016/j.trac.2008.06.002
29. Nishizawa M, Menon VP, Martin CR (1995) Metal nanotubule membranes with electrochemically switchable ion-transport selectivity. Science 268(5211):700–702. doi:10.1126/science.268.5211.700
30. Vlassiouk I, Siwy ZS (2007) Nanofluidic diode. Nano Lett 7(3):552–556. doi:10.1021/nl062924b
31. Ali M, Yameen B, Neumann R, Ensinger W, Knoll W, Azzaroni O (2008) Biosensing and supramolecular bioconjugation in single conical polymer nanochannels. Facile incorporation of biorecognition elements into nanoconfined geometries. J Am Chem Soc 130(48):16351–16357. doi:10.1021/ja8071258
32. Rothman JE, Lenard J (1977) Membrane asymmetry. Science 195(4280):743–753
33. Shaw RS, Packard N, Schroter M, Swinney HL (2007) Geometry-induced asymmetric diffusion. Proc Natl Acad Sci USA 104(23):9580–9584. doi:10.1073/pnas.0703280104
34. Kalman EB, Sudre O, Vlassiouk I, Siwy ZS (2009) Control of ionic transport through gated single conical nanopores. Anal Bioanal Chem 394(2):413–419. doi:10.1007/s00216-008-2545-3
35. Hou X, Dong H, Zhu DB, Jiang L (2010) Fabrication of stable single nanochannels with controllable ionic rectification. Small 6 (3):361–365. doi: 10.1002/smll.200901701
36. Ito Y, Park YS (2000) Signal-responsive gating of porous membranes by polymer brushes. Polym Adv Technol 11(3):136–144. doi:10.1002/1099-1581(200003)11:3<136:aid-pat961>3.0.co;2-f
37. Apel P (2001) Track etching technique in membrane technology. Radiat Meas 34(1–6):559–566. doi:10.1016/s1350-4487(01)00228-1
38. Kalman EB, Vlassiouk I, Siwy ZS (2008) Nanofluidic bipolar transistors. Adv Mater 20(2):293–297. doi:10.1002/adma.200701867
39. Xie Y, Wang X, Xue J, Jin K, Chen L, Wang Y (2008) Electric energy generation in single track-etched nanopores. Appl Phys Lett 93(16). doi:16311 10.1063/1.3001590
40. Pan YV, Wesley RA, Luginbuhl R, Denton DD, Ratner BD (2001) Plasma polymerized *N*-isopropylacrylamide: synthesis and characterization of a smart thermally responsive coating. Biomacromolecules 2(1):32–36. doi:10.1021/bm0000642
41. Sun TL, Wang GJ, Feng L, Liu BQ, Ma YM, Jiang L, Zhu DB (2004) Reversible switching between superhydrophilicity and superhydrophobicity. Angew Chem Int Edit 43(3):357–360. doi:10.1002/anie.200352565

42. Lee JW, Kim SY, Kim SS, Lee YM, Lee KH, Kim SJ (1999) Synthesis and characteristics of interpenetrating polymer network hydrogel composed of chitosan and poly (acrylic acid). J Appl Polym Sci 73(1):113–120. doi:10.1002/(sici)1097-4628(19990705)73:1<113:aid-app13>3.3.co;2-4
43. Siwy ZS (2006) Ion-current rectification in nanopores and nanotubes with broken symmetry. Adv Funct Mater 16(6):735–746. doi:10.1002/adfm.200500471
44. Kasianowicz JJ, Brandin E, Branton D, Deamer DW (1996) Characterization of individual polynucleotide molecules using a membrane channel. Proc Natl Acad Sci USA 93(24):13770–13773. doi:10.1073/pnas.93.24.13770
45. Sowerby SJ, Broom MF, Petersen GB (2007) Dynamically resizable nanometre-scale apertures for molecular sensing. Sensor Actuat B-Chem 123(1):325–330. doi:10.1016/j.snb.2006.08.031
46. Branton D, Deamer DW, Marziali A, Bayley H, Benner SA, Butler T, Di Ventra M, Garaj S, Hibbs A, Huang X, Jovanovich SB, Krstic PS, Lindsay S, Ling XS, Mastrangelo CH, Meller A, Oliver JS, Pershin YV, Ramsey JM, Riehn R, Soni GV, Tabard-Cossa V, Wanunu M, Wiggin M, Schloss JA (2008) The potential and challenges of nanopore sequencing. Nat Biotechnol 26(10):1146–1153. doi:10.1038/nbt.1495
47. Xia F, Jiang L (2008) Bio-inspired, smart, multiscale interfacial materials. Adv Mater 20(15):2842–2858. doi:10.1002/adma.200800836

Chapter 5
Asymmetric Conical Shaped Single Composite Nanochannel Materials

5.1 Introduction

The fabrication of artificial nanochannels is becoming the focus of attention. Because compared with their biological counterparts, they offer flexibility in terms of shape, size, and surface properties for real-world applications. Here we introduce an approach, which is not subject to the solution environment restriction, to prepare the stable and controllable continuous change of ionic current rectification of the asymmetric composite nanochannels using ion sputtering technology [1].

At present, there are several approaches to build single nanopores/nanochannels, such as ion-beam sculpting of Si_3N_4 membranes [2], Si/SiO_2 membrane shrinking in transmission electron microscopy [3, 4], electrochemical etching of glass membranes [5], and chemical etching of single-track polymer membranes [6]. Because the nanochannel is small enough for interactions between the channel surface and chemical species in the solution, the chemical modification of the nanochannel surface confers great flexibility in developing functional nanochannel materials [7]. Even though several solution methods [8–12] of coating nanochannels with functional molecules, which enable the size and ion-transport properties of the nanochannels to be easily tuned, have been developed during the past few years, how to endow nanochannels with asymmetric chemical properties is still a challenging task.

The chemical properties and chemical modification of artificial nanochannels are critical components [13] in the advance of smart functional nanochannels that are tunable by ambient stimuli, such as specific ions [14], applied force [15], light [5], and pH [16]. Recently, the ionic current rectification of a single nanochannel prepared by the ion-track-etching technique has been studied in comparison with the properties of various ion channels [17]. In those studies, the surface properties of the nanochannel wall are significant for the development of the desired functionality of bio-inspired nanochannel apparatus, such as biosensors [18, 19], nanofluidic devices [20, 21], and biomimetic nanochannel systems [11, 14, 16].

Inspired by asymmetrical ion channels, including both the various components of ion channels that are not of uniform distribution and the structural asymmetry,

X. Hou, *Bio-inspired Asymmetric Design and Building of Biomimetic Smart Single Nanochannels*, Springer Theses,
DOI: 10.1007/978-3-642-38050-1_5, © Springer-Verlag Berlin Heidelberg 2013

the generation of artificial ion channels [22] has strong implications for simulating the different processes of ion transport as well as enhancing the functionality of biological ion channels. Herein, we report a novel approach to develop the controllable continuous change of ionic current rectification of the nanochannel by ion sputtering technology. This is a simple preparation method to obtain stable (more than 2 months), controllable channel size, and chemically/structurally asymmetric artificial single nanochannel materials with different ionic current rectification and to produce promising materials for the development of advanced functionalized nanochannels. This innovative approach for improving metal–polymer composite asymmetric nanochannels can be considered as a novel platform to develop the methods for the asymmetric chemical modification of nanochannels that other polymer or metal–polymer nanochannels cannot achieve, such as electroless deposition [23], chemical modification in solution [10], electrostatic self-assembly [11], and contactless electrofunctionalization [12] to cover the whole inner surface of the nanochannels.

A variety of approaches [6, 24–26] have been used to prepare and study the influences of the different shapes of conical nanochannels. In this work, we focus on the influence of Pt deposition, which brings both chemical and structural asymmetry. Figure 5.1a illustrates our experimental design. We prepared a single nanochannel membrane with the well-developed ion-track-etching technique. The nanochannel was embedded in a PET membrane with a single ion track in the center. The track etching technique allowed control over the shape of the channels, and in this work the etched single nanochannel is conelike. Diameter measurements of both sides of the conical nanochannels were estimated using atomic force microscopy (AFM) method (Fig. 5.2). The large opening (base) is usually controlled from ~400 to ~2,500 nm, and the small opening (tip) is from ~2 to ~700 nm. Because of the 12 μm thickness, the high length/diameter ratio contributes to the observed lengthy translocations, effectively providing a higher temporal resolution for the application of DNA sequencing [27]. After the single nanochannel was created, each side of the nanochannel was independently sputtered to control the channel size for studying the ionic current rectification properties (Fig. 5.1a, paths 1 and 2, and Fig. 5.3). After that, we also studied the properties of double-side sputtering of the nanochannel (Fig. 5.1a, paths 3 and 4). Figure 5.1b shows scanning electron microscopy (SEM) images of the tip of the nanochannels, indicating the size of the channels for different sputtering times. The base of a nanochannel after ion sputtering is also shown in Fig. 5.1c.

5.2 Materials and Methods

5.2.1 Materials

See Tables 5.1, 5.2.

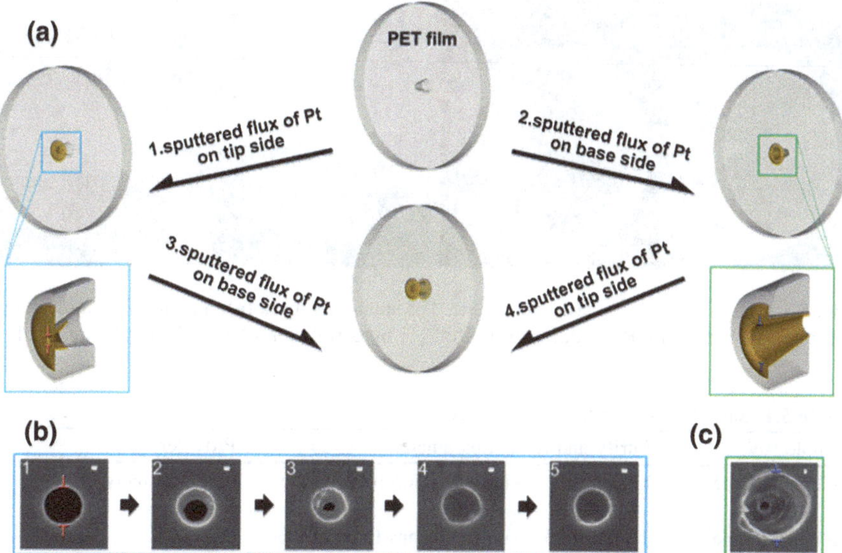

Fig. 5.1 Metal–polymer composite asymmetric nanochannel. **a** Schemeof the experimental design: (*1*) Pt was sputtered on the tip side of the conical nanochannel; (*2*) Pt was sputtered on the base side of the conical nanochannel; (*3* and *4*) Pt was sputtered on the other side of the conical nanochannels (drawing not to scale). **b** SEM images of the tip side of the nanochannels, indicating the size of the channels for different sputtering times (sputtering current I_s = 2 mA, sputtering time of the tip side t_{tip1-5} = 10–90 s). Scale bars: 100 nm. **c** SEM image of the base side of a nanochannel after ion sputtering (I_s = 2 mA, sputtering time of the base side t_{base} = 90 s). Scale bar: 100 nm. Reproduced from Ref. [1] by permission of John Wiley & Sons Ltd

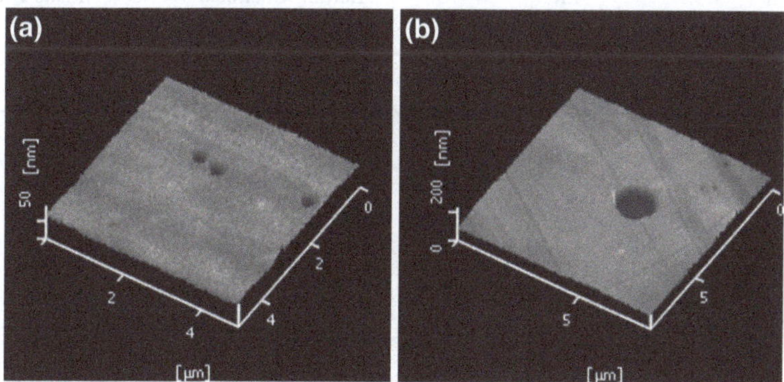

Fig. 5.2 The AFM of the conical nanochannels before ion sputtering. **a** Tip. **b** Base. Reproduced from Ref. [1] by permission of John Wiley & Sons Ltd

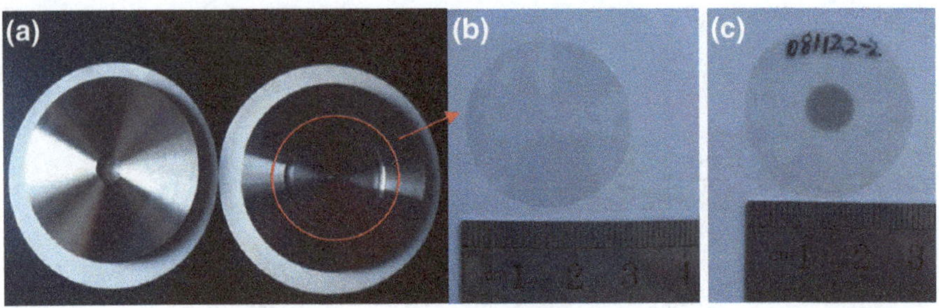

Fig. 5.3 The photos of the mask and sample. The sputtering mask (**a**), the sample before (**b**) and after ion sputtering (**c**). Reproduced from Ref. [1] by permission of John Wiley & Sons Ltd

Table 5.1 Samples used in the experiments

Sample	Purity and related parameters	Provider
PET membrane	12 μm thick, with single ion track in the center	GSI, Germany
PET membrane	12 μm thick, with multi-ions track in the center ($10^7/cm^2$)	
PET membrane	12 μm thick, without any treatment	Hostaphan RN12 Hoechst, Germany

Table 5.2 Reagents used in the experiments

Reagent	Purity quotient	Provider
KCl	A.R.	Beijing Chemical Works
HCl	A.R.	Beijing Chemical Works (contents 36–38 %)
NaOH	A.R.	Beijing Chemical Works
KOH	A.R.	West Long Chemical Co., Ltd., China
HCOOH	A.R.	Beijing Chemical Works (contents > 88 %)
Platinum sheet	99.99 %	Made in China
Filamentary silver	99.99 %	Made in China
Platinum filament	99.99 %	Made in China

5.2.2 Instruments

See Table 5.3.

5.2.3 Solvent Preparation

Track etching solution for the nanochannel preparation: 9 M NaOH
Stopping solution for the etching solution: 1 M KCl + 1 M HCOOH
Transmembrane current test solution: KCl (0.1–2 M) and (1 M, pH 2.3–10.7)

Table 5.3 Instruments used in the experiments

Instrument	Producer
Water purification system (water, 18.2 MΩ)	Milli-Q, Millipore Corporation, USA
KEITHLEY 6,487 picoammeter	Keithley Instruments, Cleveland, OH, USA
Atomic force microscope (AFM) SPI3800-SPA-400	Seiko Instruments Inc., Japan
OCA20, contact angle (CA) measuring instrument	Dataphysics, Germany
Ion sputtering instrument, SBC-12	TYKY Technology Co., Ltd., Beijing
ZF-7 Uviol lamp	Shanghai Gucun Optic Instrument Factory, China
Model PHS-25 pH meter instruction	Shanghai Precision & Scientific Instrument Co., Ltd.
Field emission scanning electron microscope, JSM-6700F	JEOL, Japan
X-ray photoelectron spectroscopy, ESCALab220i-XL	Thermo VG Scientific Ltd., UK
Homemade PTFE electrolyzer	ICCAS, China
Homemade plexiglass electrolyzer	ICCAS, China
Hot plate clarkson H3400-HS07	IKA, USA
Stopwatch timer	Made in China
Electronic thermometre	Made in China

5.3 Experiment Operation

5.3.1 Sample Preparation

The single conical nanochannel investigated here was produced in polymer films using the ion-track-etching technique. Before the chemical etching process, the samples of PET were exposed to the UV light for 1 h from each side. To produce a conical nanochannel, etching was performed only from one side, the other side of the cell contains a solution that is able to neutralize the etchant as soon as the channel opens, thus slowing down the further etching process. To additionally stop the etching, the voltage (1 V) used to monitor the etching process was applied in such a way that the negatively charged ions of the etchant were drawn out of the channel tip. The following are the etching and stopping solutions for the etching of PET: 9 M NaOH for etching, 1 M KCl + 1 M HCOOH for stopping. The large opening of the conical nanochannel was called base, while the small opening was called tip. The diameter of the base and tip was estimated from the multitrack membranes etch rate measured in the parallel etching experiments with AFM (Fig. 5.2), due to the difficulty of locating both the tip and base opening in AFM. In this work, its base is usually controlled from ~400 to ~2,500 nm, and its tip is from ~2 to ~700 nm. Pt films were deposited on the samples by ion sputtering with a Pt target, using an ion sputtering system (SBC-12) in a vacuum. The vacuum before switching on the ion source was 4 Pa and the working temperature was 25 °C. In the experimental setup, the sputtering current was kept constant (2 mA) and the deposited amount of Pt was adjusted by changing the sputtering time. The photos of the sputtering mask and the sample before and after ion sputtering are given in Fig. 5.3.

5.3.2 AFM Imaging

The single conical nanochannel investigated here was produced in polymer films using the ion-track-etching technique. The diameter of the base and tip was estimated from the multitrack membranes etch rate measured in parallel etching experiments by AFM, due to the difficulty of locating both the tip and base opening in AFM.

5.3.3 Current Measurement

The ion transport properties of the nanochannel were studied by measuring ion current through the nanochannels before and after the ion sputtering. The ion current was measured by a Keithley 6,487 picoammeter. A single-channel PET membrane was mounted between two chambers of the etching cell mentioned above. Ag/AgCl electrodes were used to apply a transmembrane potential across the film. Forward voltage was the potential applied on the base side. The main transmembrane potential used in this work was evaluated (Fig. 5.4) and a scanning voltage varied from −2 to +2 V with a 40 s period was selected. The concentration and pH of the electrolyte were also evaluated (Fig. 5.5), and we finally chose 1 M potassium chloride solution as electrolyte. Each test was repeated five times to obtain the average current value at different voltages. The testing temperature was 20 °C.

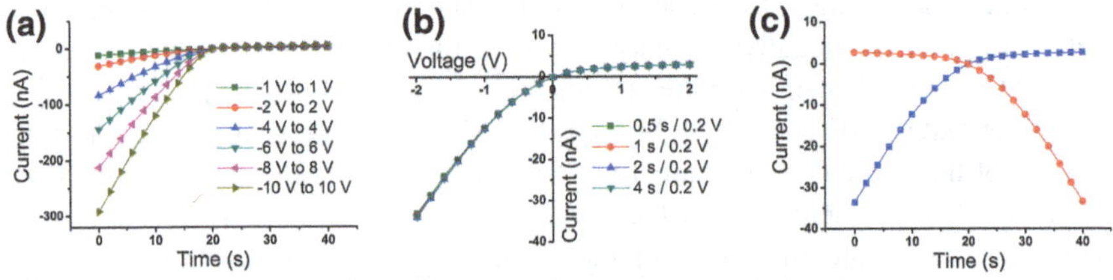

Fig. 5.4 The different current measurement conditions of the transmembrane potential across the single nanochannel. **a** The different scanning voltages varied from 1 to 10 V with a 40 s period, the voltage is 0 V at 20th s. **b** The different step time of the scanning voltage (2 V). **c** The different directions of the scanning voltage (2 V). Reproduced from Ref. [1] by permission of John Wiley & Sons Ltd

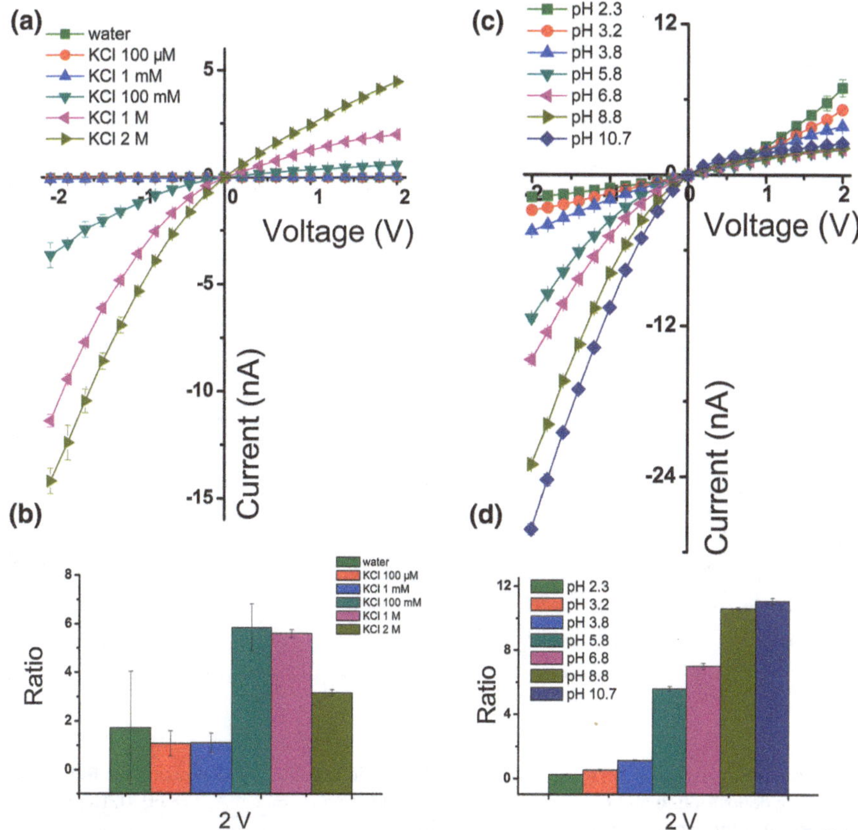

Fig. 5.5 The different concentrations of KCl solutions for the current measurements of the single nanochannel. **a** *I–V* curves of the different KCl concentrations. **b** The current ratio in the single nanochannel with different KCl concentrations at a potential of 2 V. The influence of different pH values on the current measurements of the single nanochannel (KCl 1 M). **c** *I–V* curves of the different pH conditions. **d** The current ratio in the single nanochannel with different pH at a potential of 2 V. Reproduced from Ref. [1] by permission of John Wiley & Sons Ltd

5.3.4 XPS Testing

XPS measurements were performed to validate the chemical identity of the PET films before and after ion sputtering deposition. XPS data were obtained by an ESCALab220i-XL electron spectrometer from VG Scientific using 300 W Al Kα radiation, and the base pressure was about 3×10^{-9} mbar. The binding energies were referenced to the C1s line at 284.8 eV from adventitious carbon. We observed the reduction of the C1s-O1s ratio after ion sputtering deposition (Fig. 5.6a, b). This shows that the increasing oxygen may come from the oxygen bonding with platinum to form metallic oxide compounds, used to stabilize the chemical/structure asymmetric artificial single nanochannel materials. Figure 5.6c

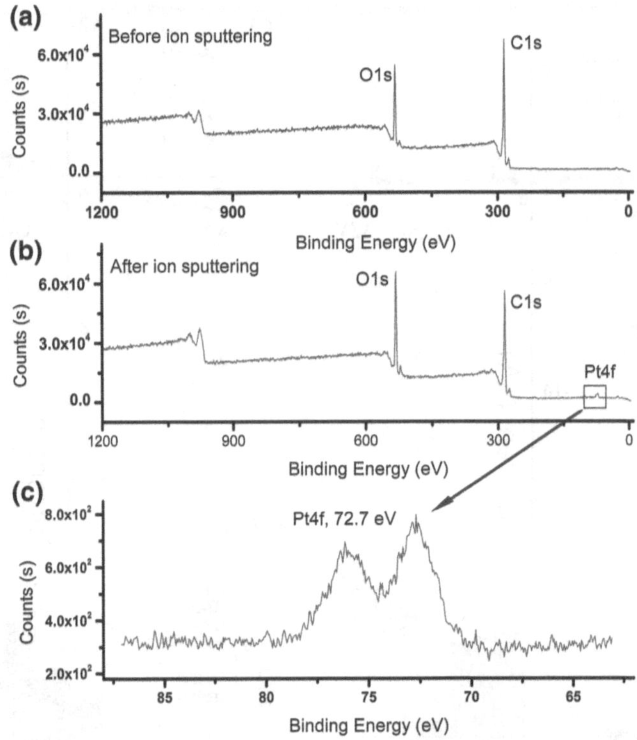

Fig. 5.6 XPS spectra of PET films before and after ion sputtering deposition. **a** Before ion sputtering deposition. **b** and **c** After ion sputtering deposition ($I_s = 2$ mA, $t = 60$ s). Reproduced from Ref. [1] by permission of John Wiley & Sons Ltd

shows that the binding energy of the Pt4f7/2 is 72.7 eV. According to standard data of XPS spectroscopy [28] the binding energy of the nonvalent Pt4f7/2 is 70.9 eV. This may indicate that Pt of nanochannel materials is at oxidation state. Moreover, as we know the ray depth of XPS is generally 2–10 nm, according to the XPS results (Tables 5.4, 5.5, 5.6, 5.7), it shows that the tip side of the depth ($I_s = 2$ mA, $t = 60$ s) is less than 10 nm.

5.3.5 CA Measurement

Contact angles were measured using an OCA20 CA system at ambient temperature and saturated humidity. The original sample was treated with 9 M NaOH for 1 h. The sample was then taken out from the etching solution and treated with the stopping solution (1 M KCl + 1 M HCOOH) for half an hour. After that, the sample was treated with the deionized water, and had been stored for 1 day in the deionized water before further experiments. Before the CA test, the sample was

Table 5.4 The XPS data from the PET film before ion sputtering; the angle of incidence is 90°

Name	Start BE	Peak BE	End BE	Height counts	FWHM eV	Area (P) CPS eV	At. %	SF
C1s, 284.8 eV	291.77	284.87	280.67	15,225.28	1.6	33898.14	77.65	1
O1s, 532.2 eV	537.9	533.32	527.62	7636.97	3.03	24872.49	22.35	2.93

Reproduced from Ref. [1] by permission of John Wiley & Sons Ltd

Table 5.5 The XPS data from the PET film after ion sputtering; the angle of incidence is 15°

Name	Start BE	Peak BE	End BE	Height counts	FWHM eV	Area (P) CPS eV	At. %	SF
C1s, 284.8 eV	292.7	284.74	279.4	4,966.23	1.87	13,881.15	74.06	1
O1s, 532.0 eV	537.86	532.25	526.93	3,761.39	3.1	12,226.59	25.57	2.93
Pt4f, 72.7 eV	81.25	72.8	67.21	251.78	1.94	1,192	0.37	15.46

Reproduced from Ref. [1] by permission of John Wiley & Sons Ltd

Table 5.6 The XPS data from the PET film after ion sputtering; the angle of incidence is 45°

Name	Start BE	Peak BE	End BE	Height counts	FWHM eV	Area (P) CPS eV	At. %	SF
C1s, 284.8 eV	292.51	284.83	280.14	8,151.9	1.86	21,785.63	72.93	1
O1s, 532.2 eV	537.9	532.22	527.65	6,763.93	2.93	20,410.17	26.78	2.93
Pt4f, 72.7 eV	81.42	72.94	68	341.88	2.05	1,513.78	0.3	15.46

Reproduced from Ref. [1] by permission of John Wiley & Sons Ltd

Table 5.7 The XPS data from the PET film after ion sputtering; the angle of incidence is 90°

Name	Start BE	Peak BE	End BE	Height counts	FWHM eV	Area (P) CPS eV	At. %	SF
C1s, 284.8 eV	292.62	284.79	280.72	12,306.63	1.91	33,369.65	71.96	1
O1s, 532.2 eV	538.32	532.02	527	10,395.39	1.87	32,866.64	27.78	2.93
Pt4f, 72.7 eV	82.2	72.71	67.05	461.76	2.41	2,055.02	0.26	15.46

Reproduced from Ref. [1] by permission of John Wiley & Sons Ltd

blown dry by N_2. Deionized water droplets (about 2 μl) were then dropped carefully onto the surfaces. The average CA value was obtained at five different positions of the same sample.

The CAs of the PET film surfaces after ion sputtering (Fig. 5.7b) were larger than that of the one before ion sputtering (Fig. 5.7a). The CA results indicate that the ion sputtering deposition could lead to a remarkable change of the wettability of the surface (from 68.1 ± 1.6° to 88.2 ± 1.2°), which means the change of the chemical composition. These results may potentially spark the further development of experimental approach to study the relationship between the wettability and the ion transport property on confined space in nanoscale.

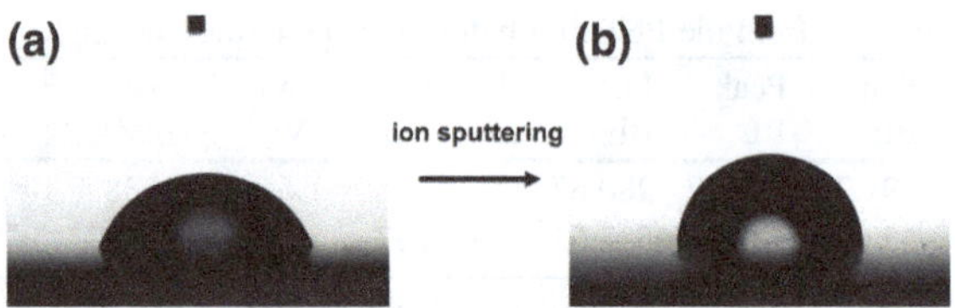

Fig. 5.7 The CA of the PET film surfaces before and after ion sputtering. **a** Before ion sputtering. **b** After ion sputtering ($I_s = 2$ mA, $t = 60$ s). Reproduced from Ref. [1] by permission of John Wiley & Sons Ltd

5.4 Results and Discussion

The ion transport properties of the nanochannel (Sample1) before and after ion sputtering have been examined by current measurements. The results indicate that, before ion sputtering, the nanochannel exhibits a linear *I–V* curve, which means it does not rectify (Fig. 5.8a). It is worth mentioning that there was a remarkable

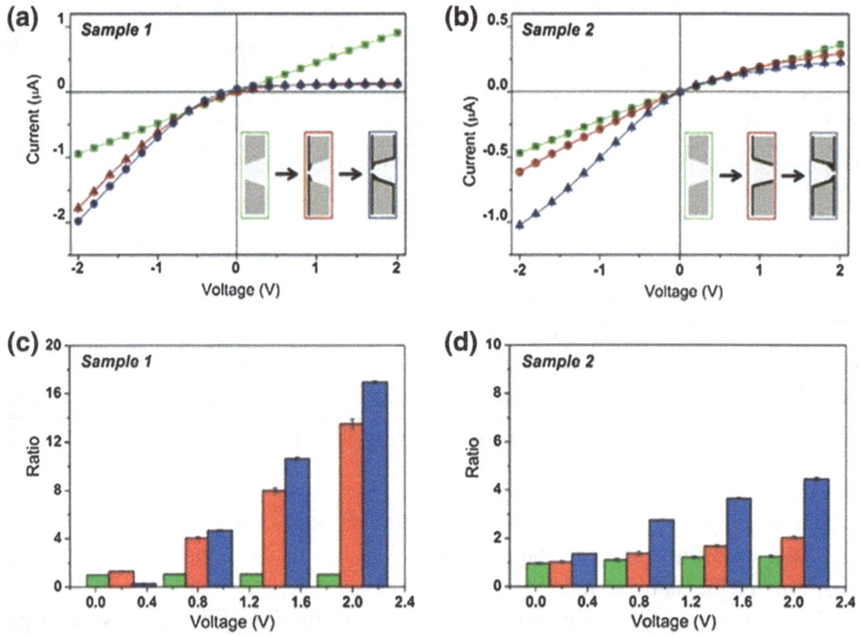

Fig. 5.8 Ion transport properties of the nanochannels before and after ion sputtering. **a, b** *I–V* properties of the nanochannels before (*green*) and after ion sputtering (*red* and *blue*). Sample1: $I_s = 2$ mA, $t_{tip} = 110$ s, $t_{base} = 30$ s; Sample2: $I_s = 2$ mA, $t_{base} = 110$ s, $t_{tip} = 30$ s. The sputtering sequence is illustrated in the insets of (**a**) (from ▲ tip to ● base) and (**b**) (from ● base to ▲ tip). **c**, **d** Voltage-dependent behavior of currents of the nanochannels before (*green*) and after ion sputtering (*red* and *blue*) at 0.2, 0.8, 1.4, and 2 V for Samples1 and 2, respectively. Reproduced from Ref. [1] by permission of John Wiley & Sons Ltd

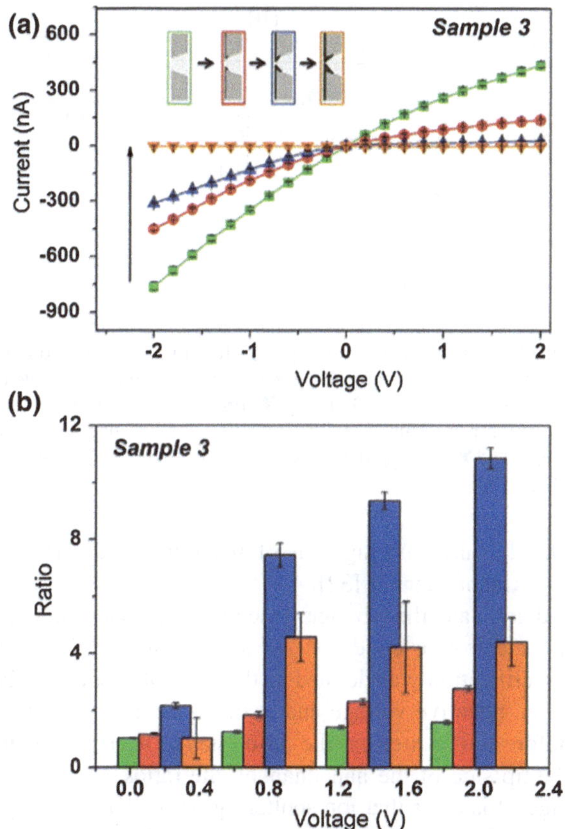

Fig. 5.9 Dependence of the ionic-current rectification on different sputtering times. **a** *I–V* properties of the nanochannel (Sample3) before (*green*) and after ion sputtering on the tip side with different sputtering times. The sputtering sequence is illustrated in the inset; ■ before, ● 1st, ▲ 2nd, ▼ 3rd. $I_s = 2$ mA, $t_{tip,1st} = 30$ s (*red*), $t_{tip,2nd} = 30$ s (*blue*), $t_{tip,3rd} = 30$ s (*orange*). **b** Voltage-dependent behavior of currents of the nanochannel before (*green*) and after ion sputtering (*red*, *blue*, and *orange*) at 0.2, 0.8, 1.4, and 2 V. Reproduced from Ref. [1] by permission of John Wiley & Sons Ltd

difference after ion sputtering on the tip side of the nanochannel. A significant ionic rectification was observed as asymmetric *I–V* curves. Then, Pt was sputtered on the base side of the nanochannel and there was very clear rectification. Figure 5.8b shows that the *I–V* properties of the nanochannel (Sample2) are in a different order. Obviously, these *I–V* properties also exhibit the asymmetric property after ion sputtering, but ion sputtering on different sides of the nanochannel makes a visible difference to the rectification. We observed that the key factor to obtain the rectification property is ion sputtering on the tip side of the nanochannel. This novel asymmetric nanochannel may provide a new experimental model for complete theoretical studies [26, 29], molecular dynamics

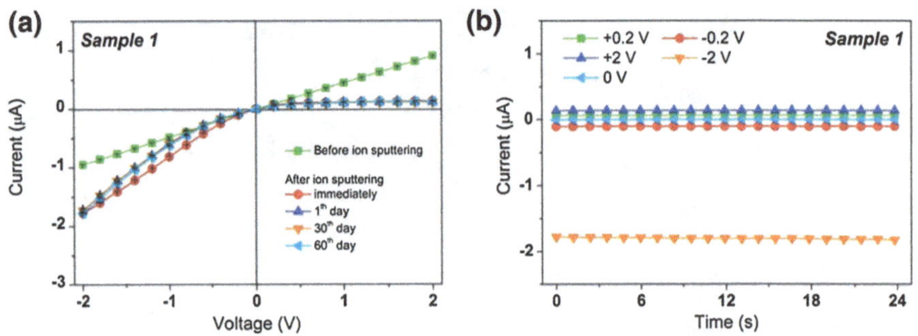

Fig. 5.10 Stability of the nanochannel (Sample1) after ion sputtering on the tip side. **a** *I–V* properties of the nanochannel before (■, *green*) and after ion sputtering on different days (immediately ●, *red*; 1st day ▲, *blue*; 30th day ▼, *orange*; 60th day ◀, *sky blue*). **b** Constant-potential properties of the nanochannel at different voltages (30th day: 0.2 V ■, *green*; −0.2 V ●, *red*; 2 V ▲, *blue*; −2 V ▼, *orange*; 0 V ◀, *sky blue*). Reproduced from Ref. [1] by permission of John Wiley & Sons Ltd

simulations [30, 31], and investigation of the effects of surface properties on nanoscale flow in a nanochannel [32].

Figure 5.8c, d indicate the voltage-dependent behavior of currents recorded under symmetric electrolyte conditions (1 M potassium chloride solution). The degree of ionic rectification was defined as the ratio of absolute values of currents recorded at a given negative voltage and at the same absolute value of a positive voltage [33]. Before ion sputtering, the ratios stayed nearly unchanged. After ion sputtering on the tip side of the nanochannel, the ratios markedly increased with increasing voltage. It is clear that ion sputtering on the base of the nanochannel is not the conspicuous factor for this phenomenon, as shown in Fig. 5.8d.

According to the fact that ion sputtering on the tip of the nanochannel is the key factor to obtain the rectification property, we examined the dependence of the ionic current rectification on different sputtering times. Figure 5.9a shows that the *I–V* properties of the nanochannel are in the following order: before ion sputtering (green curve); the first time of ion sputtering on the tip side of the nanochannel (red curve); the second time (blue curve); and the third time (orange curve). It is clear that a different ionic current rectification of the nanochannel was obtained by controlling the tip channel sizes with different sputtering times. There was a maximum ratio during this process, as shown in Fig. 5.9b. It was assumed that the influential factors of the change in ratio were determined by both the chemical and structural asymmetry [32, 34, 35]. The results indicated that sputtering enhances the rectification, but after the third sputtering, the very small tip size made the ion transport current value very close to the detection limit of the testing equipment. Thus, the testing voltage-dependent behavior of the current was not a true reflection of the rectification of the nanochannel with the present experimental apparatus. We are convinced that, with the further development of ion sputtering in this system, the channel size of the nanochannel and the ionic rectification will be precisely modulated.

We also investigated the stability of the nanochannels. As shown in Fig. 5.10a, they can be stabilized for more than 2 months. Figure 5.10b also shows the very stable measurement currents at different constant voltages, which is an important feature for application in sensors. To further explore the surface properties and the stability of the PET films before and after ion sputtering, the films were also studied by CA measurements and XPS analysis (Figs. 5.6 and 5.7). The CA results indicate that ion sputtering deposition could lead to a remarkable change of the wettability of the surface (from 68.1 ± 1.6° to 88.2 ± 1.2°), which means a change of the chemical composition. XPS shows that the Pt of the nanochannel materials might be in the oxidized state, due to oxygen bonding with platinum to form metallic oxide compounds. Therefore, these stable nanochannels are appropriate for further development of sensor systems, such as DNA sequencing [36–38].

5.5 Conclusions

In summary, we have demonstrated experimentally a novel approach to prepare stable and chemically/structurally asymmetric artificial single nanochannels with different ionic current rectification by ion sputtering technology. This method achieves a wide range of control to obtain a channel size from hundreds of nanometers to a very small size in accordance with the different dimension requirements of various kinds of sensors. This ionic current rectification property is controlled by ion sputtering deposition, which brings both chemical and structural asymmetry to the nanochannels. Such nanochannel materials could potentially spark further experimental and theoretical efforts with the choice of a different metal, such as Au, for further asymmetric chemical modification with complicated functional molecules for the exploitation of more complex smart nanochannel materials. Moreover, they may provide a new experimental model for complete theoretical studies, molecular dynamics simulations, and investigation of the effects of surface properties on nanoscale flow in a nanochannel.

References

1. Hou X, Dong H, Zhu DB, Jiang L (2010) Fabrication of stable single nanochannels with controllable ionic rectification. Small 6(3):361–365. doi:10.1002/smll.200901701
2. Li J, Stein D, McMullan C, Branton D, Aziz MJ, Golovchenko JA (2001) Ion-beam sculpting at nanometre length scales. Nature 412(6843):166–169. doi:10.1038/35084037
3. Storm AJ, Chen JH, Ling XS, Zandbergen HW, Dekker C (2003) Fabrication of solid-state nanopores with single-nanometre precision. Nat Mater 2(8):537–540. doi:10.1038/nmat941
4. Iqbal SM, Akin D, Bashir R (2007) Solid-state nanopore channels with DNA selectivity. Nat Nanotechnol 2(4):243–248. doi:10.1038/nnano.2007.78
5. Wang G, Bohaty AK, Zharov I, White HS (2006) Photon gated transport at the glass nanopore electrode. J Am Chem Soc 128(41):13553–13558. doi:10.1021/ja064274j

6. Harrell CC, Siwy ZS, Martin CR (2006) Conical nanopore membranes: controlling the nanopore shape. Small 2(2):194–198. doi:10.1002/smll.200500196
7. Gyurcsanyi RE (2008) Chemically-modified nanopores for sensing. Trac-Trends Anal Chem 27(7):627–639. doi:10.1016/j.trac.2008.06.002
8. Harrell CC, Kohli P, Siwy Z, Martin CR (2004) DNA—nanotube artificial ion channels. J Am Chem Soc 126(48):15646–15647. doi:10.1021/ja044948v
9. Nilsson J, Lee JRI, Ratto TV, Letant SE (2006) Localized functionalization of single nanopores. Adv Mater 18(4):427–431. doi:10.1002/adma.200501991
10. Vlassiouk I, Siwy ZS (2007) Nanofluidic diode. Nano Lett 7(3):552–556. doi:10.1021/nl062924b
11. Ali M, Yameen B, Neumann R, Ensinger W, Knoll W, Azzaroni O (2008) Biosensing and supramolecular bioconjugation in single conical polymer nanochannels. facile incorporation of biorecognition elements into nanoconfined geometries. J Am Chem Soc 130(48):16351–16357. doi:10.1021/ja8071258
12. Bouchet A, Descamps E, Mailley P, Livache T, Chatelain F, Haguet V (2009) Contactless electrofunctionalization of a single pore. Small 5(20):2297–2303. doi:10.1002/smll.200900482
13. Baker LA, Bird SP (2008) Nanopores—a makeover for membranes. Nat Nanotechnol 3(2):73–74. doi:10.1038/nnano.2008.13
14. Hou X, Guo W, Xia F, Nie FQ, Dong H, Tian Y, Wen LP, Wang L, Cao LX, Yang Y, Xue JM, Song YL, Wang YG, Liu DS, Jiang L (2009) A biomimetic potassium responsive nanochannel: G-Quadruplex DNA conformational switching in a synthetic nanopore. J Am Chem Soc 131(22):7800–7805. doi:10.1021/Ja901574c
15. Huh D, Mills KL, Zhu X, Burns MA, Thouless MD, Takayama S (2007) Tuneable elastomeric nanochannels for nanofluidic manipulation. Nat Mater 6(6):424–428. doi:10.1038/nmat1907
16. Xia F, Guo W, Mao YD, Hou X, Xue JM, Xia HW, Wang L, Song YL, Ji H, Qi OY, Wang YG, Jiang L (2008) Gating of single synthetic nanopores by proton-driven DNA molecular motors. J Am Chem Soc 130(26):8345–8350. doi:10.1021/Ja800266p
17. Siwy Z, Apel P, Baur D, Dobrev DD, Korchev YE, Neumann R, Spohr R, Trautmann C, Voss KO (2003) Preparation of synthetic nanopores with transport properties analogous to biological channels. Surf Sci 532:1061–1066. doi:10.1016/s0039-6028(03)00448-5
18. Martin CR, Siwy ZS (2007) Learning nature's way: biosensing with synthetic nanopores. Science 317(5836):331–332. doi:10.1126/science.1146126
19. Wharton JE, Jin P, Sexton LT, Horne LP, Sherrill SA, Mino WK, Martin CR (2007) A method for reproducibly preparing synthetic nanopores for resistive-pulse biosensors. Small 3(8):1424–1430. doi:10.1002/smll.200700106
20. Jirage KB, Hulteen JC, Martin CR (1997) Nanotubule-based molecular-filtration membranes. Science 278(5338):655–658. doi:10.1126/science.278.5338.655
21. Kalman EB, Vlassiouk I, Siwy ZS (2008) Nanofluidic bipolar transistors. Adv Mater 20(2):293–297. doi:10.1002/adma.200701867
22. Gracheva ME, Melnikov DV, Leburton J-P (2008) Multilayered semiconductor membranes for nanopore ionic conductance modulation. ACS Nano 2(11):2349–2355. doi:10.1021/nn8004679
23. Nishizawa M, Menon VP, Martin CR (1995) Metal nanotubule membranes with electrochemically switchable ion-transport selectivity. Science 268(5211):700–702. doi:10.1126/science.268.5211.700
24. Scopece P, Baker LA, Ugo P, Martin CR (2006) Conical nanopore membranes: solvent shaping of nanopores. Nanotechnology 17(15):3951–3956. doi:10.1088/0957-4484/17/15/057
25. Xu F, Wharton JE, Martin CR (2007) Temptate synthesis of carbon nanotubes with diamond-shaped cross sections. Small 3(10):1718–1722. doi:10.1002/smll.200700306
26. Ramirez P, Apel PY, Cervera J, Mafe S (2008) Pore structure and function of synthetic nanopores with fixed charges: tip shape and rectification properties. Nanotechnology 19(31):315707. doi:10.1088/0957-4484/19/31/315707

27. Mara A, Siwy Z, Trautmann C, Wan J, Kamme F (2004) An asymmetric polymer nanopore for single molecule detection. Nano Lett 4(3):497–501. doi:10.1021/nl035141o
28. Wagner CD, Muilenberg GE et al (1979) Handbook of x-ray photoelectron spectroscopy: a reference book of standard data for use in x-ray photoelectron spectroscopy. Eden Prairie, Minn: Physical Electronics Division, Perkin-Elmer Corp, ©1979
29. Letant SE, Schaldach CM, Johnson MR, Sawvel A, Bourcier WL, Wilson WD (2006) Pore conductivity control at the hundred-nanometer scale: an experimental and theoretical study. Small 2(12):1504–1510. doi:10.1002/smll.200600263
30. Stefureac RI, Lee JS (2008) Nanopore analysis of the folding of zinc fingers. Small 4(10):1646–1650. doi:10.1002/smll.200800585
31. Kalra A, Garde S, Hummer G (2003) Osmotic water transport through carbon nanotube membranes. Proc Natl Acad Sci USA 100(18):10175–10180. doi:10.1073/pnas.1633354100
32. Yang SC (2006) Effects of surface roughness and interface wettability on nanoscale flow in a nanochannel. Microfluid Nanofluid 2(6):501–511. doi:10.1007/s10404-006-0096-5
33. Siwy ZS (2006) Ion-current rectification in nanopores and nanotubes with broken symmetry. Adv Funct Mater 16(6):735–746. doi:10.1002/adfm.200500471
34. Vlassiouk I, Smirnov S, Siwy Z (2008) Nanofluidic ionic diodes. comparison of analytical and numerical solutions. Acs Nano 2(8):1589–1602. doi: 10.1021/nn800306u
35. Karnik R, Duan C, Castelino K, Daiguji H, Majumdar A (2007) Rectification of ionic current in a nanofluidic diode. Nano Lett 7(3):547–551. doi:10.1021/nl062806o
36. Yan H, Xu BQ (2006) Towards rapid DNA sequencing: detecting single-stranded DNA with a solid-state nanopore. Small 2(3):310–312. doi:10.1002/smll.200500464
37. Dekker C (2007) Solid-state nanopores. Nat Nanotechnol 2(4):209–215. doi:10.1038/nnano.2007.27
38. Healy K, Schiedt B, Morrison AP (2007) Solid-state nanopore technologies for nanopore-based DNA analysis. Nanomedicine 2(6):875–897. doi:10.2217/17435889.2.6.875

Zeitfracht Medien GmbH
Ferdinand-Jühlke-Straße 7
99095 Erfurt, Deutschland
produktsicherheit@kolibri360.de